Integrated Circuits and Systems

Series Editor
Anantha Chandrakasan
Massachusetts Institute of Technology
Cambridge, Massachusetts

For further volumes, go to
http://www.springer.com/series/7236

Masashi Horiguchi • Kiyoo Itoh

Nanoscale Memory Repair

Springer

Dr. Masashi Horiguchi
Renesas Electronics Corporation
5-20-1, Josuihon-cho
Kodaira-shi, Tokyo, 187-8588
Japan
masashi.horiguchi.kc@renesas.com

Dr. Kiyoo Itoh
Hitachi Ltd.
Central Research Laboratory
1-280, Higashi-Koigakubo
Kokubunji-shi, Tokyo, 185-8601
Japan
kiyoo.itoh.pt@hitachi.com

ISSN 1558-9412
ISBN 978-1-4614-2794-0 ISBN 978-1-4419-7958-2 (eBook)
DOI 10.1007/978-1-4419-7958-2
Springer New York Dordrecht Heidelberg London

Printed on acid-free paper

Springer is part of Springer Science+Business Media (www.springer.com)

Preface

Repair techniques for nanoscale memories are becoming more important to cope with ever-increasing "errors" causing degraded yield and reliability. In fact, without repair techniques, even modern CMOS LSIs, such as MPUs/SoCs, in which memories have dominated the area and performances, could not have been successfully designed. Indeed, various kinds of errors have been prominent with larger capacity, smaller feature size, and lower voltage operations of such LSIs. The errors are categorized as hard/soft errors, timing/voltage margin errors, and speed-relevant errors. Hard/soft errors and timing/voltage margin errors, which occur in a chip, are prominent in a memory array because the array comprises memory cells having the smallest size and largest circuit count in the chip. In particular, coping with the margin errors is vital for low-voltage nanoscale LSIs, since the errors rapidly increase with device and voltage scaling. Increase in operating voltage is one of the best ways to tackle the issue. However, this approach is unacceptable due to intolerably increased power dissipation, calling for other solutions by means of devices and circuits. Speed-relevant errors, which are prominent at a lower voltage operation, comprise speed-degradation errors of the chip itself and intolerably wide chip-to-chip speed-variation errors caused by the ever-larger interdie design-parameter variation. They must also be solved with innovative devices and circuits. For the LSI industry, in order to flourish and proliferate, the problems must be solved based on in-depth investigation of the errors.

Despite the importance, there are few authoritative books on repair techniques because the solutions to the problems lie across different fields, e.g., mathematics and engineering, logic and memories, and circuits and devices. This book systematically describes the issues, based on the authors' long careers in developing memories and low-voltage CMOS circuits. This book is intended for both students and engineers who are interested in the yield, reliability, and low-voltage operation of nanoscale memories. Moreover, it is instructive not only to memory designers, but also to all digital and mixed-signal LSI designers who are at the leading edge of such LSI developments.

Chapter 1 describes the basics of repair techniques. First, after categorizing sources of hard/soft errors, the reductions by means of redundancy, error checking and correction (ECC), and their combination are comprehensively described. Second, after defining the minimum operating voltage V_{min}, reductions of timing/voltage margin errors are described in terms of V_{min}. Finally, reduction techniques for speed-relevant errors are briefly discussed.

Chapter 2 deals with a detailed explanation of the redundancy techniques for repairing hard errors (faults), where faulty memory cells are replaced by spare memory cells provided on the chip in advance. Various yield models and calculations are introduced and various practical circuits and architectures that the authors regard as important for higher yield and reliability are discussed. The chapter also describes the devices for memorizing the addresses of faults and testing techniques for redundancy.

Chapter 3 describes the details of the ECC techniques to cope with both hard and soft errors, where extra bits (check bits) are added to original data bits, thereby enabling error detection and/or correction. After mathematical preparations, various error-correcting codes used for the techniques and their practical implementations in various memory LSIs are discussed. This is followed by the estimation of the reduction in hard-error and soft-error rates using ECC. Testing techniques for ECC are also described.

Chapter 4 deals with the combination of the redundancy and ECC. Combining both the techniques generates a synergistic effect and dramatically enhances the repair capability. It is especially effective for random-bit errors. After quantitative estimation of the synergistic effect, the application to the repair of faults due to device mismatch is discussed as a promising application of the effect.

Chapter 5 systematically describes challenges to ultra-low-voltage nanoscale memories and the repair techniques to accomplish the issues. After clarifying that reduction in the minimum operating voltage V_{DD} (i.e., V_{min}) is the key to reducing the voltage and timing margin error, adaptive circuits and relevant technologies to reduce V_{min} are proposed, and the general features are described. Then, the V_{min}s of logic gates, SRAMs, and DRAMs are compared. After that, devices (e.g., fully depleted planar SOI and FinFET structures), circuits (e.g., gate-source reverse biasing schemes accepting low threshold voltage (V_t) MOSFETs), and subsystems to widen the margins through reducing V_{min} are described.

Chapter 6 briefly describes device/circuit techniques to cope with two kinds of speed-relevant errors, namely, the speed degradation error and the interdie speed variation error. After specifying reduced gate-over-drive voltage of MOSFETs as the source of the speed degradation error, some solutions (e.g., using low-V_{t0} circuits and dynamic V_t circuits utilizing double-gate FD-SOI structures) are exemplified. Moreover, after specifying the so-called global variation of design parameters in the wafer as the source of the interdie speed variation error, some solutions such as power management for compensating for the variation with static or quasi-static controls of internal supply voltages are presented.

Tokyo, Japan

Masashi Horiguchi
Kiyoo Itoh

Contents

Chapter 1
An Introduction to Repair Techniques

1.1 Introduction

With larger capacity, smaller feature size, and lower voltage operations of memory-rich CMOS LSIs (Fig. 1.1), various kinds of "errors (faults)" have been prominent and the repair techniques for them have become more important. The "errors" are categorized as hard/soft errors, timing/voltage margin errors, and speed-relevant errors. Hard/soft errors and timing/voltage margin errors, which occur in a chip, are prominent in a memory array because the array comprises memory cells having the smallest size and largest circuit count in the chip. In particular, coping with the margin errors is becoming increasingly important, and thus an emerging issue for low-voltage nanoscale LSIs, since the errors rapidly increase with device and voltage scaling. Increase in operating voltage is one of the best ways to tackle the issue. However, this approach is not acceptable due to intolerably increased power dissipation. Speed-relevant errors, which are prominent at a lower voltage operation, include speed-degradation errors of the chip itself and intolerably wide chip-to-chip speed-variation errors caused by the ever-larger interdie design-parameter variation. For the LSI industry in order to flourish and proliferate, solutions based on in-depth investigation of the errors are crucial.

In this chapter, first, after categorizing sources of hard/soft errors, reductions of hard/soft errors by means of redundancy, error checking and correction (ECC), and their combination are comprehensively described. Second, after defining the minimum operating voltage V_{min}, reductions of timing/voltage margin errors are described in terms of V_{min}. Finally, reduction techniques for speed-relevant errors are briefly discussed.

1.2 Hard and Soft Errors and Repair Techniques

There are many sources of hard and soft errors. To repair the errors, redundancy and on-chip ECC, as shown in Fig. 1.2, and the combination are effective.

M. Horiguchi and K. Itoh, *Nanoscale Memory Repair*, Integrated Circuits and Systems 1,
DOI 10.1007/978-1-4419-7958-2_1,

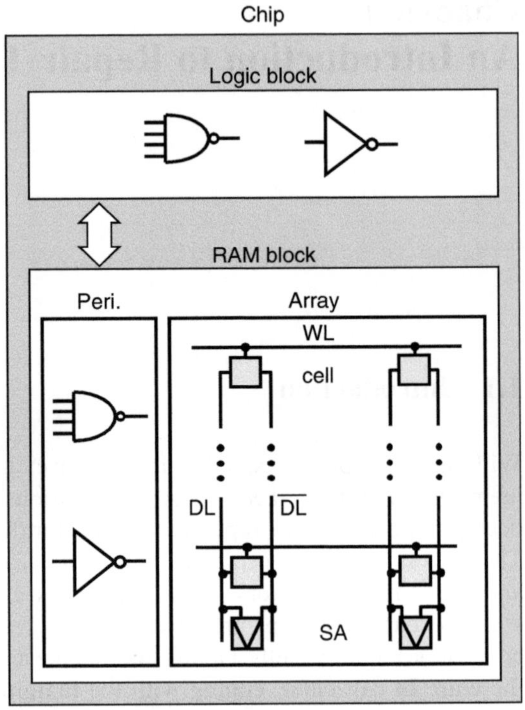

Fig. 1.1 LSI chip composed of logic block and RAM block. Peri. peripheral logic circuits, DL, $\overline{DL}$ data (bit) lines, and SA sense amplifier

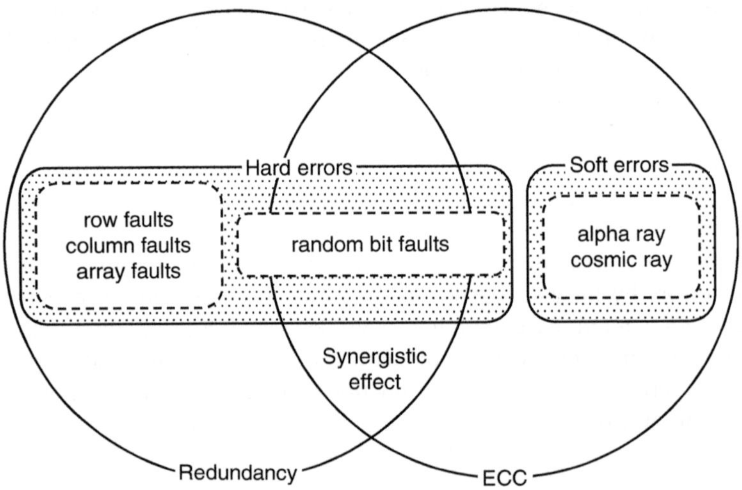

Fig. 1.2 Repair techniques for memories: redundancy and error checking and correction (ECC)

1.2.1 Hard and Soft Errors

The hard errors, which are mainly caused by various fabrication-process defects, permanently cause incorrect operations of a part of an LSI chip or an entire chip.

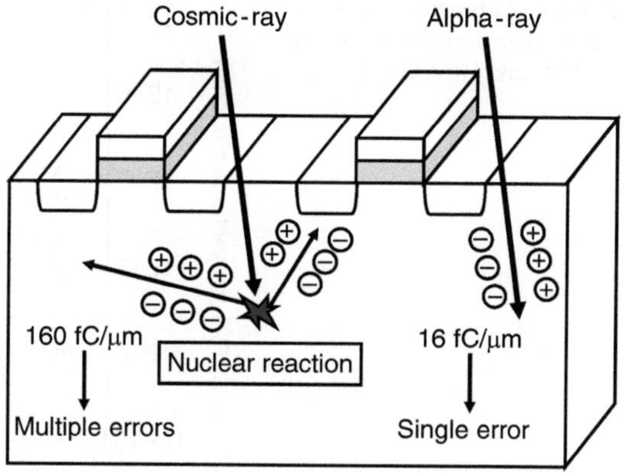

Fig. 1.3 Soft errors. Reproduced from [4] with permission; © 2010 IEEE

They are classified into single-bit faults (only a memory cell is faulty), row/column faults (all the memory cells on a row/column are faulty), array faults (all the memory cells in a memory array are faulty), etc., according to the number of faulty cells and their distribution. The hard errors increase with device scaling due to need for more complicated process and device technologies.

The soft errors are phenomena that the information stored in memory cell nodes and other circuit nodes is lost. They are caused by noises, especially by the incidents of alpha ray or cosmic ray. If the charge generated by the incidents is collected at the node that retains information, an upset of the node voltage may occur (Fig. 1.3) [1–4]. Since devices are not broken by soft errors, correct operations are possible after rewriting the lost information. The smaller the signal charge of the node (i.e., product of the voltage and capacitance), the larger the soft error rate (SER). Hence, SER increases with device and voltage scaling because signal charge is reduced, and the SER of memory cells usually having the smallest signal charge are thus larger than that of logic gates. In particular, the soft-error problem is problematic for SRAMs with device scaling, as seen in Fig. 1.4. The rapidly reduced signal charge of the cell is responsible for the increase because a capacitor is not added intentionally at stored nodes, unlike dynamic random-access memory (DRAM) cells. Even for logic circuits, alpha ray or cosmic ray causes soft errors.

1.2.2 Redundancy

The redundancy technique is used for repairing hard errors. The idea was proposed in the 1960s [5, 6] and was applied to actual 64–256-kb RAMs in the late 1970s [7–9]. Since then, it has been widely used as effective methods of enhancing production yield and reducing cost per bit of DRAMs. In this scheme, faulty memory cells are replaced by spare memory cells, which are provided on the chip in advance. The replacement is controlled by programmable devices, in which the addresses of faulty memory cells are programmed before shipping. Usually, a row/column of memory

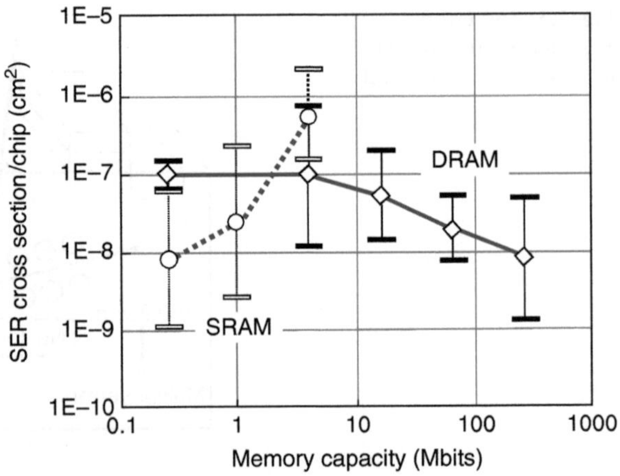

Fig. 1.4 Comparisons of soft-error characteristics between SRAMs and DRAMs [7]

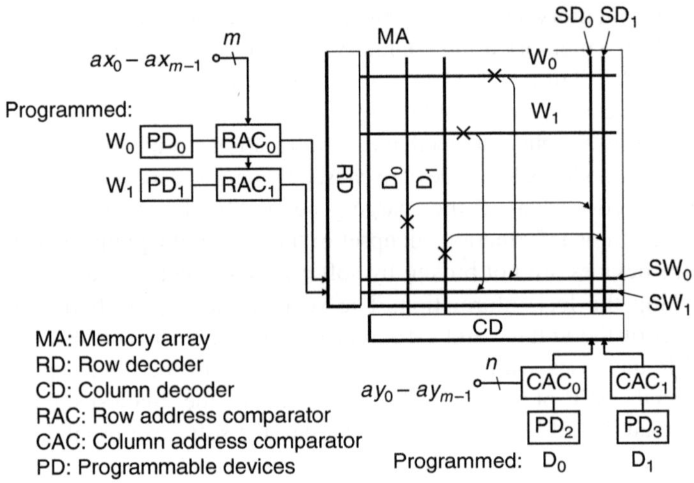

Fig. 1.5 Principle of redundancy for a RAM [7]

cells (usually wordlines and/or datalines), including faulty ones is replaced by a spare row/column because the memory cells are arranged in a matrix. Figure 1.5 shows a well-known redundancy technique [7] applied to a RAM. Here, the memory has two spare wordlines SW_0, SW_1 and the same number of row-address comparators RAC_0, RAC_1, and programmable devices PD_0, PD_1. If the addresses of the faulty wordlines (W_0, W_1) have been programmed into the devices during wafer testing, the address comparators allow one of the spare wordlines (SW_0, SW_1) to be activated whenever a set of input row-address signals ($ax_0 - ax_{m-1}$) during actual operations matches one of the faulty addresses. Thus, W_0 and W_1 are replaced by SW_0 and SW_1 (SW_0 and SW_1 are used instead of W_0 and W_1), respectively. Similarly, faulty data lines D_0 and D_1 are replaced by spare datalines SD_0 and SD_1, respectively. In modern RAMs, as many as over 100 faulty elements could be replaced by spare elements in the early stage of production, with an additional chip area of less than 5%. The programmable devices are usually poly-Si fuses, which are blown by means of a laser beam or a pulsed current, although they do accept antifuses or nonvolatile memory cells [7]. Laser programming occupies a smaller chip area and does not normally affect circuit performance, but it does require special test equipment and increases wafer handling and testing time. Also, the laser spot size and beam-positioning requirements become more stringent for ever finer line widths. On the other hand, electrical programming by a pulsed current is carried out using standard test equipment. Usually, a hole is cut in the passivation glass over such fuses to reduce the amount of programming current needed. The possibility of mobile-ion contamination of active circuit areas can be eliminated by using guard-ring structures surrounding the fuse area, or other techniques [7]. The area and performance penalties of electrical programming can be minimized by careful circuit design. Electrical programming is used when the number of fuses required is not large enough to offset the negative aspects of laser programming. In any event, laser programming has been widely accepted due to the small area and performance penalties, simplicity, assurance of cutting, and the ease of laying out fuses.

1.2.3 ECC

In the past, off-chip ECC technique was used for enhancing the reliability of main storages of computer systems. Nowadays, on-chip ECC technique has been widely used in various memory LSIs, since the importance was recognized by the discovery of soft error caused by alpha ray [1]. On-chip ECC is effective for correcting both soft errors and hard errors (faults) of RAMs [10–12]. The ECC technique provides extra memory cells on a chip to enable ECC using the information stored in the cells (check bits). Unlike the redundancy, programmable devices to store faulty addresses are not necessary. The principle of ECC is shown in Fig. 1.6. During write operation, the coding circuit generates check bits from the input data bits. The generation rule is determined by the error-correcting code used. The set composed of the data bits and the check bits generated according to the rule is called a *code word*. The code word is written into the memory array so that the stored information has a certain amount of redundancy. If there is a fault in the memory array or a soft error occurs before reading, the read data contain one or more errors. However, the error(s) can be detected and corrected by the decoding circuit if the number of errors does not exceed the correction capability of the error-correcting code. The operation of the decoding circuit is as follows. First, check bits are generated from the read data bits just like the coding circuit and are compared with the read check bits. If no error exists, both are the same. If they are not the same, the position(s) of erroneous bit(s) are detected by analyzing the comparison results and the errors are corrected before outputting. The corrected data are usually written back to the memory array to prevent the accumulation of soft errors. In DRAMs, checking and correcting are performed during every refresh operation [10] as well as during read operation.

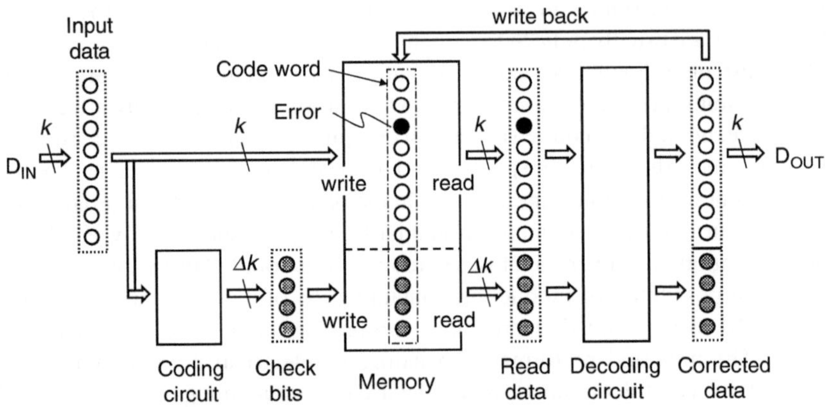

Fig. 1.6 Principle of on-chip ECC for a RAM [7]

Table 1.1 Number of required check bits for single-error correction (SEC) code

Number of data bits k	4	8	16	32	64	128	256
Number of check bits Δk	3	4	5	6	7	8	9

The key to designing an on-chip ECC circuit is the selection of a suitable error-correcting code. Error-correcting codes are classified into single-error correction (SEC) codes, single-error correction and double-error detection (SEC–DED) codes, double-error correction (DEC) codes, etc., according to the error-detection/correction capability. Table 1.1 shows the number of required check bits of a SEC code, Δk, for k data bits. It should be noted that the ratio $\Delta k/k$ decreases as k increases. Although a large k reduces the memory-area penalty, it results in both large circuit-area and access-time penalties. This is because the number of logic gates for the encoding and decoding circuits is approximately proportional to $k \log_2 k$ and the number of stages of the circuits is proportional to $\log_2 k$. Therefore, the design of an ECC circuit requires a compromise among the memory area, circuit area, and access time. The error-correcting codes that have been applied to RAMs until now include Hamming codes [12], extended Hamming codes [11], and bidirectional parity codes [10]. The former one is an SEC code, while the latter two are SEC–DED codes. DEC codes were not realistic for on-chip ECC because of considerable enlargement of the encoding/decoding circuits. For nanoscale memories, however, a DEC (or more) code might be necessary because of the increase in error rates, while the circuit-area penalty is reduced due to device scaling [13]. The correcting efficiency is closely related to array architectures as well as device structures immune to soft errors, as explained in Chaps. 3 and 5.

Comparison Between Redundancy and ECC: Both the redundancy and the ECC have their own advantages and disadvantages. Therefore, they are used for repairing different types of errors as shown in Fig. 1.2. The differences are summarized as follows.

(1) The addresses of faults are programmed on chip and the programmable devices to store the addresses are needed in the redundancy. On the other hand, no addresses are programmed in the ECC. The ECC can repair errors at any addresses as long as the number of errors does not exceed the correcting capability, while the redundancy can repair only the programmed addresses. In addition, the ECC can repair both soft errors and hard errors, while the redundancy can repair only hard errors.
(2) The chip-size and access-time penalties due to ECC are usually larger than those due to redundancy. The chip-size penalty includes memory cells for check bits and coding/decoding circuits, and the access-time penalty is the coding/decoding times. In addition, the number of extra cells for the ECC is determined by the error-correcting code, while that for the redundancy can be determined more freely according to the error rate and desired yield.
(3) The redundancy and the ECC are suitable for different types of hard errors (faults) as shown in Fig. 1.7. Among various types of faults in memory LSIs, row faults and column faults are effectively repaired by the redundancy (Fig. 1.7a, b). However, repairing efficiency of random-bit faults by the redundancy is not high (Fig. 1.7c). For example, random-bit faults include insufficient voltage margin (e.g., static noise margin) due to device mismatch in

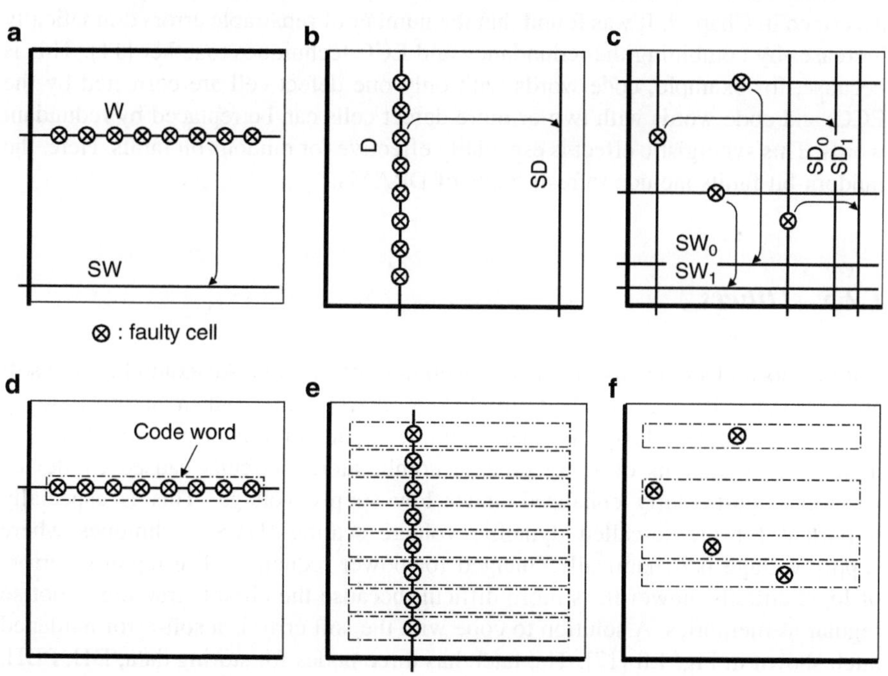

Fig. 1.7 Hard-error repair using redundancy and ECC: (**a**) a faulty row is replaced by a spare row, (**b**) a faulty column is replaced by a spare column, (**c**) the repairing efficiency of random-bit faults by redundancy is not high, (**d**) a row fault cannot be repaired by ECC, (**e**) a column fault can be theoretically repaired by ECC, and (**f**) random-bit faults can be effectively repaired by ECC

SRAM cells [14, 15], and insufficient data-retention time in DRAM cells. Even if there is only one faulty cell in a row (column), an entire row (column) of spare cells are required. Generally, random-bit faults require as many spare lines (rows or columns) as the number of faulty cells unless two or more faulty cells are located on a row (column). The ECC can effectively repair these faults if the distribution of faulty cells is random as shown in Fig. 1.7f. On the other hand, the ECC is not suitable for repairing row and column faults. Row faults cannot be repaired by ECC. A code word of ECC is usually composed of memory cells in a row because they must be simultaneously read out. Therefore a row fault causes all the bits in a code word to be faulty, making it uncorrectable (Fig. 1.7d). Column faults can be theoretically repaired by the ECC because each code word along the faulty column contains only one faulty bit. However, another faulty bit in these code words cannot be repaired (Fig. 1.7e). Therefore, the redundancy is more suitable for repairing row and column faults.

1.2.4 Combination of Redundancy and ECC

Combination of the redundancy and the ECC maximizes the repair efficiency, as described in Chap. 4. It was found that the number of repairable errors dramatically increases by combining the redundancy and ECC techniques together [11]. This is because, for example, code words with only one defect cell are corrected by the ECC, and code words with two or more defect cells can be replaced by redundant words. This synergistic effect is especially effective for random-bit faults. Here, the random-bit faults include refresh faults of DRAMs.

1.2.5 Others

Hard errors and soft errors may occur even in logic circuits. An example is the soft error that the data in a latch is lost by the charge generated in Si substrate by cosmic-ray incident [16]. Another example is the timing error caused by the increase in variations of delay time of combinatorial circuits caused by device variations, particularly enhanced under low supply voltage. This is especially important for the so-called dynamic voltage scaling (DVS) techniques where supply voltage is dynamically changed for power reduction. The repair of errors of logic circuits, however, is more difficult because the circuit structure is not so regular as memories. A solution to cope with the soft error is a soft-error hardened latch shown in Fig. 1.8 [17]. The latch has three nodes for storing data, DH, PDH, and NDH. The latter two are redundant nodes. Even if data in one of the nodes is lost by soft error, it can be recovered by using the data in other two nodes. This is a kind of ECC technique. A solution to cope with the timing error is to add another latch for capturing a delayed output of the combinatorial circuit [18–20], as shown

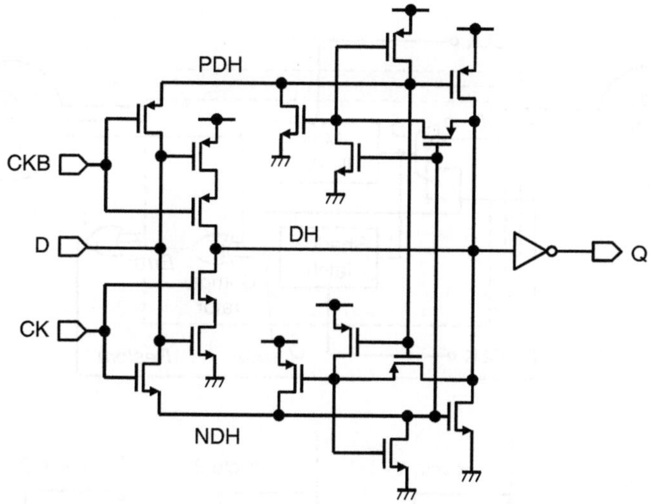

Fig. 1.8 Soft-error hardened latch. Reproduced from [17] with permission; © 2010 IEEE

in Fig. 1.9a. The main data stream is the path from logic stage 1 to logic stage 2 through the main flip-flop. The shadow latch is provided for capturing the output D of logic stage 1. Since the latch captures D after the main flip-flop, a timing error is detected by comparing the outputs of the main flip-flop and the shadow latch, Q and Q_{shadow}. Figure 1.9b shows the operating waveforms. During cycle 1, the data input *D_in* arrives before the rising edge of the clock CLK and both the main flip-flop and the shadow latch can capture the correct data. During cycle 2, however, *D_in* arrives after the rising edge of CLK and the main flip-flop cannot capture it. On the other hand, since the shadow latch is "through" state when CLK is at high level, its output Q_{shadow} alters as soon as the arrival of *D_in*. Since the outputs Q and Q_{shadow} do not match, the error signal is asserted. The content of the shadow latch is transferred to the main flip-flop during the next cycle and Q is corrected. Thus, correct operations can be resumed. This is a kind of ECC technique using duplicate circuits.

1.3 Margin Errors and Repair Techniques

The margin error of nanoscale LSIs results from reduced timing and voltage margins. It is closely related to operating voltage V_{DD}, variations of process, voltage, and temperature (i.e., PVT variations), and the speed target. Operating voltage V_{DD} must be higher than the sum of the minimum operating voltage V_{DD} (V_{min}) determined by intrinsic V_t-variation (ΔV_t) of MOSFETs, other PVT variations, and an additional voltage ΔV_τ necessary to meet the speed target. Here, other PVT variations include the power supply droop and noise ΔV_{ps} that depends on power supply integrity, and

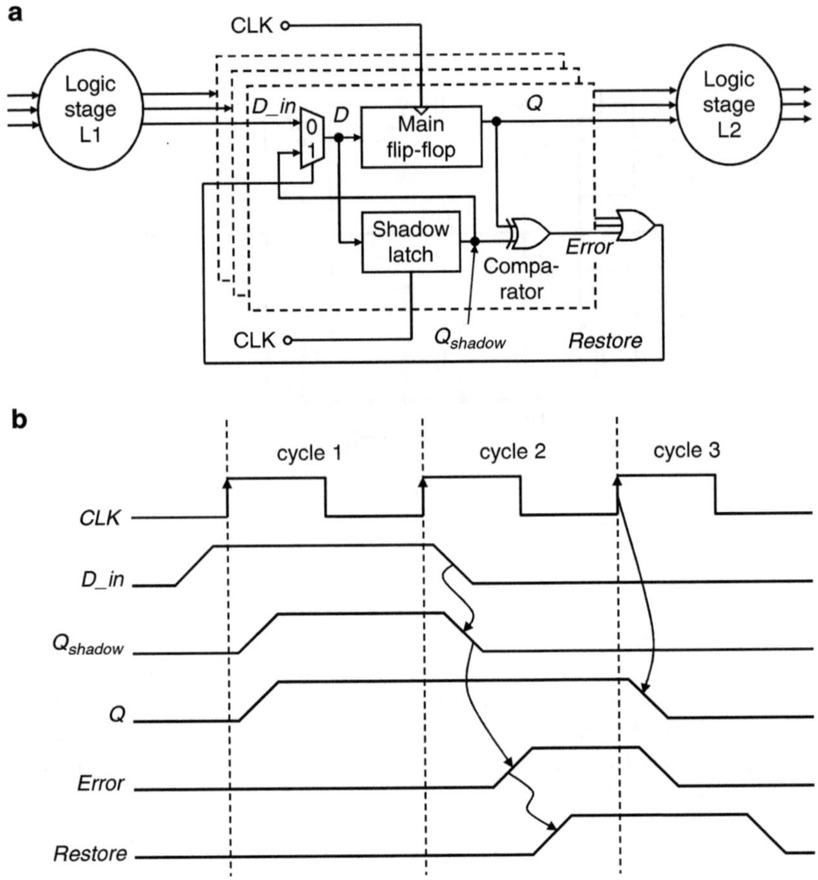

Fig. 1.9 Principle of timing-error correction of logic circuit: (**a**) circuit diagram and (**b**) timing diagrams. Reproduced from [18] with permission; © 2010 IEEE

a necessary voltage ΔV_{ext} to offset the effect imposed mainly by extrinsic V_t-variation of MOSFETs. The larger the difference between V_{DD} and the sum, the wider the margins. Since ΔV_τ cannot be specified here, reducing V_{min}, ΔV_{ps}, and ΔV_{ext} as much as possible is essential to widen the margins under a given V_{DD}. In particular, reducing V_{min} is the key because V_{min} rapidly increases with device scaling due to the ever-increasing ΔV_t, so it eventually dominates V_{DD} (see Fig. 5.1). Even so, ΔV_{ps} and ΔV_{ext} must also be reduced because they start to govern the margins when V_{min} dominates V_{DD}. ΔV_{ps} is reduced with small cores and chips, new architectures such as multicore MPUs, and the compact 3-D integration of small chips with high density through silicon vias (TSVs). A drastic reduction in the memory array area is also vital to further reduce ΔV_{ps} since the array dominates the core or chip. They ensure excellent power supply integrity and low noise in signal lines throughout the subsystem.

In the following, the V_{min} issue is mainly discussed because it is closely related to the timing and voltage margin error. Then, the interdie speed variation issue caused by ΔV_{ext} is briefly discussed.

1.3.1 Device and Process Variations

There are some process and device parameter variations [21] (Figs. 1.10 and 1.11) as sources of margin errors, which are prominent in the nanoscale era. The V_t-variation in a chip is dominated by random dopant fluctuations (RDFs) in the number and location of dopant atoms in the channel region of MOSFETs. It is called the local V_t-variation due to the chip-level variation. The local V_t-variation narrows the timing and voltage margin of circuits in a chip, and increases V_t-mismatch between a pair of MOSFETs in differential circuits such as sense amplifiers and latch circuits. On the other hand, even in the absence of RDF, there are other sources of the variations, that is, variations of the channel length (L) and its edge roughness, gate-oxide thickness, and implant uniformities of MOSFETs. They are called global or systematic variations because they are the wafer-level variations. Of the global variations, the short-channel effect (SCE) is a major concern. It presents the V_t lowering (Fig. 1.12), in which V_t decreases as the channel length L decreases, and thus giving a significant impact on the speed variation. For example, a MOSFET is slowest when having the longest L and thus the highest V_t, while fastest when having the shortest L and thus the lowest V_t. This increases speed variations between chips (i.e., causing interdie speed variations), and any chip that does not meet speed specifications is rejected as a faulty chip.

1.3.2 Timing and Voltage Margin Errors

The timing and voltage margins can comprehensively be explained on the assumption that a chip comprises many identical circuit sets. Then, characteristics of the chip

•Highly random effects
Random dopant fluctuation (RDF)
Line-edge roughness
Local oxide thickness variations
Interface charge nonuniformities

•Patterning proximity effects
Optical proximity correction (OPC)

•Proximity effects associated with stress, polish, and anneals
Overlayers, STI-induced, RTA-generated

•Device-related
Channel length
Pocket implants poly grains
Oxide thickness

•Design-related
Hot spots, Droop

Fig. 1.10 Variation components in nanoscale LSIs [21]

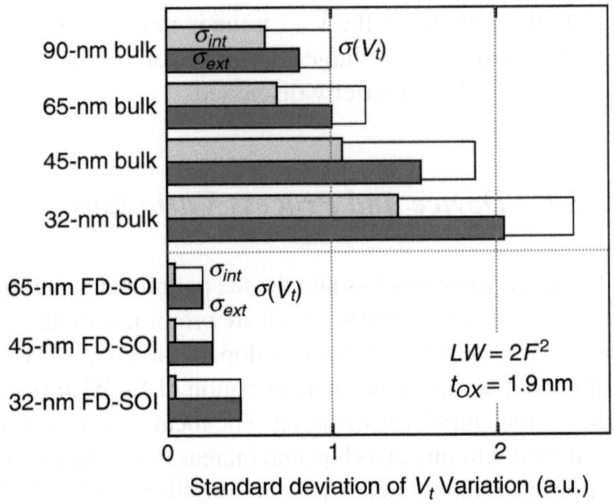

Fig. 1.11 Standard deviations of threshold voltage (V_t) variation, $\sigma(V_t)$, and intrinsic (σ_{int}) and extrinsic (σ_{ext}) V_t variations. The intrinsic variation is due to uneven dopant distribution, while the extrinsic variation is due to variations of gate length, gate width, oxide thickness, etc. Bulk and fully depleted silicon-on-insulator (FD-SOI) devices are compared. Reproduced from [15] with permission; © 2010 IEEE

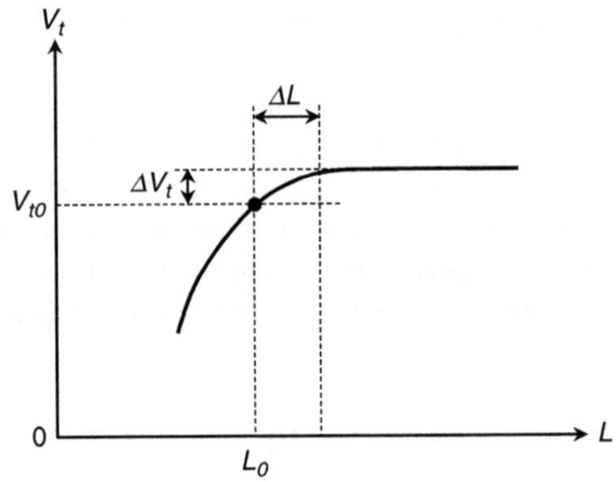

Fig. 1.12 V_t lowering due to short-channel effects

are explained by those of the circuit. Figure 1.13 illustrates a simple model of timing skew, where two identical circuits each composed of MOSFETs drive identical loads connected to the next stage. When the statistical variations in V_t in a chip are not negligible any more, the two MOSFETs in a certain set in the chip can statistically have two different V_ts as the difference of ΔV_t. Here, the discharging or charging speed $\tau(V_t)$ of a nanoscale MOSFET having a V_t is expressed as follows. The drain current of the conventional MOSFET is proportional to $(V_{DD} - V_t)^\eta$. The speed $\tau(V_t)$ is thus given by the charge ($\propto V_{DD}$) at the output divided by the drain current as

$$\tau(V_t) \propto \frac{V_{DD}}{W(V_{DD} - V_t)^\eta}, \tag{1.1}$$

where W is the channel width of MOSFET, and the value of η varies between 2 and 1, depending on the MOSFET operation either in saturation or linear region during switching [22]. The value η becomes less than 2 for the saturation drain current of

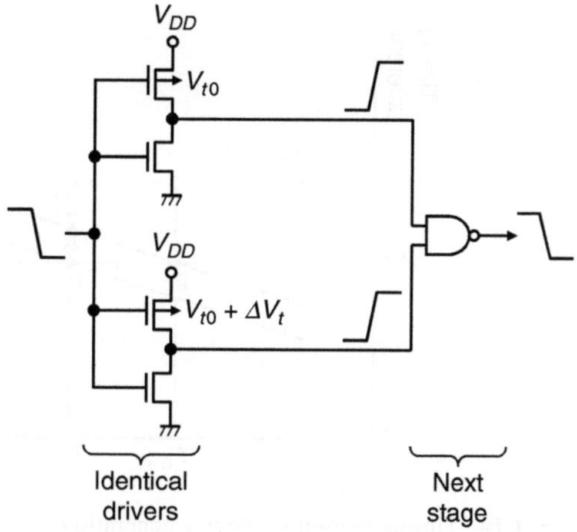

Fig. 1.13 Circuits with identical driver MOSFETs

short-channel MOSFET under velocity saturation conditions. Since the value η was 1.5 for 0.5-μm MOSFETs and 1.25 for 65-nm MOSFETs, it is assumed to be 1.2 for nanoscale MOSFETs. In any event, the two MOSFETs can have different speeds, causing a timing difference $\Delta\tau$ between the two inputs of the next stage. If the average V_t is equal to the lowest necessary average V_t (V_{t0}) that is determined by leakage specifications (see Fig. 5.5), the timing difference is given as

$$\Delta\tau = \frac{\tau(V_{t0} + \Delta V_t)}{\tau(V_{t0})} = \left(1 - \frac{\Delta V_t}{V_{DD} - V_{t0}}\right)^{-1.2}. \tag{1.2}$$

If the tolerable maximum $\Delta\tau_0$ is defined as the $\Delta\tau$ necessary for reliable timing designs, the timing margin in the fixed device generation (F_0) is given as $\Delta\tau_0 - \Delta\tau$ (Fig. 1.14). Here, $\Delta\tau_0$ is almost constant, which is determined by the design targets and circuit configurations, as discussed in Chap. 5. The margin narrows at each successive device generation since $\Delta\tau$ increases as a result of increasing ΔV_t (see a line $p_0p'_1$ in the figure). Errors occur whenever $\Delta\tau$ exceeds $\Delta\tau_0$. To maintain the margin to the same, $\Delta\tau$ must be reduced to point p_1 with increasing ($V_{DD} - V_{t0}$) at each generation, so increased ΔV_t is offset by increasing ($V_{DD} - V_{t0}$) (see a line p_0p_1). Since V_{t0} is usually constant to maintain the sub-threshold leakage current to the same, increase in V_{DD} is eventually the key to maintain the margin in the subsequent generations.

As for the voltage margin, it is closely related to $\Delta\tau_0$. If the minimum operating (functional) V_{DD} (i.e., V_{min}) is defined as the V_{DD} necessary for $\Delta\tau_0$, it is given by solving (1.2) for V_{DD} as

$$V_{min} = V_{t0} + (1 + \gamma)\Delta V_t, \quad \gamma = 1/(\Delta\tau_0{}^{1/1.2} - 1). \tag{1.3}$$

Obviously, V_{min} increases with device scaling due to increased ΔV_t for a given $\Delta\tau_0$, as shown in Fig. 1.15 and expected from Fig. 1.14. The voltage margin at a

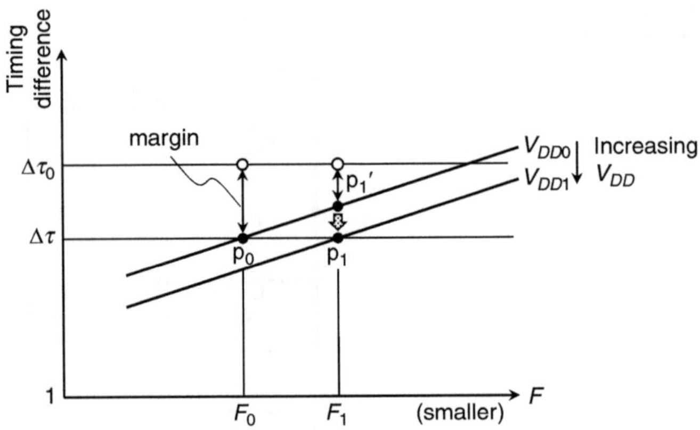

Fig. 1.14 Timing margin vs. device generation

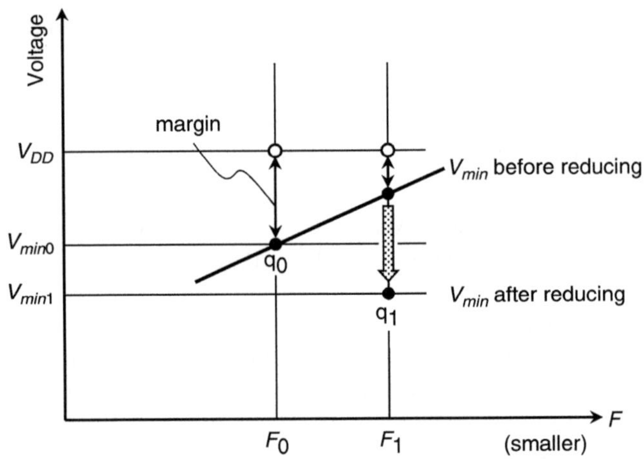

Fig. 1.15 V_{min} vs. device generation

fixed V_{DD}, given as $V_{DD} - V_{min}$, thus narrows at each successive device generation. Voltage margin errors occur whenever V_{min} exceeds V_{DD}. To maintain the margin to the same, another way different from increasing V_{DD} is necessary, which is to reduce V_{min} (point q_1 in the figure). Hence, reducing ΔV_t is strongly needed. This also contributes to widening the timing margin.

1.3.3 Reductions of Margin Errors

Repair techniques to drastically reduce margin errors are required. Using the redundancy and/or ECC techniques are effective for memories. In fact, they have considerably reduced the errors of SRAMs inherently having a larger V_t-variation. Even so, they cannot manage to reduce ever-more errors caused by the ever-increasing V_t-variation with device scaling. In addition, the techniques are ineffective for logic gates, although a few different ways have been proposed, as discussed previously. Hence, other innovative device and circuit techniques are strongly needed. Useful techniques to reduce V_t-variations and thus V_{min} are summarized as follows. The details are described in Chap. 5.

Use of Largest MOSFET Possible: It is especially effective to reduce the V_t-variation of MOSFETs in SRAM cells having the narrowest margin in a chip, as described in Chap. 5 (see Fig. 5.31b).

Dual-V_{t0} and Dual-V_{DD} Circuits: Using low-V_{t0} circuits as much as possible in a chip widen the margins because low-V_{t0} MOSFETs reduce V_t-variations, as explained in Chap. 5. The resultant leakage path, however, must be cut with the help of high-V_{t0} MOSFETs, and thus high V_{DD} to maintain a high speed. This inevitably necessitates dual-V_{t0} and dual-V_{DD} circuits. Instantaneously activated low-V_{t0} circuits backed up by high-V_{t0} circuits also reduce the variations. Increased leakage is reduced because the leakage flows only during a short period, while the output level is held in the back-up circuits without leakage, as seen in DRAM sense amplifier (see Fig. 5.40).

New MOSFETs: Advanced MOSFETs to reduce V_t-variation are needed. Good examples are using conventional MOSFETs with thinner t_{ox}, a high-k metal gate for even thinner t_{ox}, and fully depleted (FD) silicon-on-insulator (SOI) structures, such as planar ultrathin buried oxide (BOX) MOSFETs and FinFETs, for ultralow dose channel.

1.4 Speed-Relevant Errors and Repair Techniques

Two kinds of errors are more prominent with voltage and device scaling. The first comes from an excessively slow speed of an average chip. If V_{min} is reduced sufficiently, the resultant wide voltage margin enables V_{DD} to be reduced, giving a benefit of low-power dissipation. This reduction, however, is strictly limited by this kind of errors. If the resultant speed of the average chip is too slow to meet the requirement, many chips result in faults even if they correctly operate. This is due to reduced gate-over-drive of MOSFET caused by reducing V_{DD}. There are two ways to solve the problem. One is to reduce V_{t0}, even if it calls for leakage reduction circuits [7]. The other is to enlarge the channel width W to offset the degraded gate-over-drive. The second comes from the interdie speed variation. Due to an excessively wide speed variation of chips, some of the chips result in faults. Compensation for the variations with adaptive controls of internal power supply voltages is indispensable, as described in Chap. 6.

References

1. T. C. May and M. H. Woods, "Alpha-particle-induced soft errors in dynamic memories," IEEE Trans. Electron Devices, vol. ED-26, pp. 2–9, Jan. 1979.
2. K. Takeuchi, K. Shimohigashi, E. Takeda, E. Yamasaki, T. Toyabe and K. Itoh, "Alpha-particle-induced charge collection measurements for megabit DRAM cells," IEEE Trans. Electron Devices, vol. 36, pp. 1644–1650, Sep. 1989.
3. J. F. Ziegler, H. W. Curtis, H. P. Muhlfeld, C. J. Montrose, B. Chin, M. Nicewicz, C. A. Russell, W. Y. Wang, L. B. Freeman, P. Hosier, L. E. LaFave, J. L. Walsh, J. M. Orro, G. J. Unger, J. M. Ross, T. J. O'Gorman, B. Messina, T. D. Sullivan, A. J. Sykes, H. Yourke, T. A. Enger, V. Tolat, T. S. Scott, A. H. Taber, R. J. Sussman, W. A. Klein and C. W. Wahaus, "IBM experiments in soft fails in computer electronics (1978–1994)," IBM J. Res. Dev., vol. 40, pp. 3–18, Jan. 1996.
4. K. Osada, K. Yamaguchi, Y. Saitoh and T. Kawahara, "SRAM immunity to cosmic-ray-induced multierrors based on analysis of an induced parasitic bipolar effect," IEEE J. Solid-State Circuits, vol. 39, pp. 827–833, May 2004.
5. E. Tammaru and J. B. Angell, "Redundancy for LSI yield enhancement," IEEE J. Solid-State Circuits, vol. SC-2, pp. 172–182, Dec. 1967.
6. A. Chen, "Redundancy in LSI memory array," IEEE J. Solid-State Circuits, vol. SC-4, pp. 291–293, Oct. 1969.
7. K. Itoh, M. Horiguchi, and H. Tanaka, *Ultra-low voltage nano-scale memories*, Springer, New York, 2007.
8. R. P. Cenker, D. G. Clemons, W. R. Huber, J. B. Petrizzi, F. J. Procyk and G. M. Trout, "A fault-tolerant 64K dynamic RAM," ISSCC Dig. Tech. Papers, Feb. 1979, pp. 150–151.
9. R. R. DeSimone, N. M. Donofrio, B. L. Flur, R. H. Kruggel and H. H. Leung, "FET RAMs," ISSCC Dig. Tech. Papers, Feb. 1979, pp. 154–155.
10. T. Mano, J. Yamada, J. Inoue and S. Nakajima, "Circuit techniques for a VLSI memory," IEEE J. Solid-State Circuits, vol. SC-18, pp. 463–470, Oct. 1983.
11. H. L. Kalter, C. H. Stapper, J. E. Barth Jr., J. DiLorenzo, C. E. Drake, J. A. Fifield, G. A. Kelley Jr., S. C. Lewis, W. B. van der Hoeven and J. A. Yankosky, "A 50-ns 16-Mb DRAM with a 10-ns data rate and on-chip ECC," IEEE J. Solid-State Circuits, vol. 25, pp. 1118–1128, Oct. 1990.
12. K. Arimoto, K. Fujishima, Y. Matsuda, M. Tsukude, T. Oishi, W. Wakamiya, S. Satoh, M. Yamada and T. Nakano, "A 60-ns 3.3-V-only 16-Mbit DRAM with multipurpose register," IEEE J. Solid-State Circuits, vol. 24, pp. 1184–1190, Oct. 1989.
13. R. Naseer and J. Draper, "Parallel double error correcting code design to mitigate multi-bit upsets in SRAMs," Proc. ESSCIRC, Sep. 2008, pp. 222–225.
14. M. J. M. Pelgrom, A. C. J. Duinmaijer and A. P. G. Welbers, "Matching properties of MOS transistors," IEEE J. Solid-State Circuits, vol. 24, pp. 1433–1440, Oct. 1989.
15. M. Yamaoka, K. Osada, R. Tsuchiya, M. Horiuchi, S. Kimura and T. Kawahara, "Low power SRAM menu for SOC application using yin-yang-feedback memory cell technology," Symp. VLSI Circuits Dig. Tech. Papers, June 2004, pp. 288–291.
16. Y. Tosaka, S. Satoh, T. Itakura, H. Ehara, T. Ueda, G. A. Woffinden and S. A. Wender, "Measurement and analysis of neutron-induced soft errors in sub-half-micron CMOS circuits," IEEE Trans. Electron Devices, vol. 45, pp. 1453–1458, July 1998.
17. Y. Komatsu, Y. Arima, T. Fujimoto, T. Yamashita and K. Ishibashi, "A soft-error hardened latch scheme for SoC in a 90nm technology and beyond," Proc. CICC, Oct. 2004, pp. 329–332.
18. S. Das, D. Roberts, S. Lee, S. Pant, D. Blaauw, T. Austin, K. Flautner and T. Mudge, "A self-tuning DVS processor using delay-error detection and correction," IEEE J. Solid-State Circuits, vol. 41, pp. 792–804, Apr. 2006.
19. J. Tschanz, K. Bowman, S.-L. Lu, P. Aseron, M. Khellah, A. Raychowdhury, B. Geuskens, C. Tokunaga, C. Wilkerson, T. Karnik and V. De, "A 45nm resilient and adaptive

microprocessor core for dynamic variation tolerance," ISSCC Dig. Tech. Papers, Feb. 2010, pp. 282–283.
20. D. Bull, S. Das, K. Shivshankar, G. Dasika, K. Flautner and D. Blaauw, "A power-efficient 32b ARM ISA processor using timing-error detection and correction for transient-error tolerance and adaptation to PVT variation," ISSCC Dig. Tech. Papers, Feb. 2010, pp. 284–285.
21. Kelin J. Kuhn, "Reducing variation in advanced logic technologies: approaches to process and design for manufacturability of nanoscale CMOS," IEDM Proc., pp. 471–474, Dec. 2007.
22. Shih-Wei Sun and Paul G. Y. Tsui, "Limitation of CMOS supply-voltage scaling by MOSFET threshold-voltage variation," IEEE J. Solid-State Circuits, vol. 30, pp. 947–949, Aug. 1995.

Chapter 2
Redundancy

2.1 Introduction

For designing redundancy circuit, the estimation of the advantages and disadvantages is indispensable. The introduction of redundancy in a memory chip results in yield improvement and fabrication-cost reduction. However, it also causes the following penalties. First, spare memory cells to replace faulty cells, programmable devices to memorize faulty addresses, and control circuitry to increase chip size. Second, the time required for the judgment whether the input address is faulty or not is added to the access time. Third, special process steps to fabricate the programmable devices and test time to store faulty addresses into the devices are required. Therefore, the design of redundancy circuit requires a trade-off between yield improvement and these penalties. The estimation of yield improvement requires a fault-distribution model. There are two representative models, Poisson distribution model and negative-binomial model, which are often used for the yield analysis of memory LSIs. The "replacement" of normal memory elements by spare elements requires checking whether the accessed address includes faulty elements, and if yes, inhibiting the faulty element from being activated and activating a spare element instead. These procedures should be realized with as small penalty as possible. One of the major issues for the replacement is memory-array division. Memory arrays are often divided into subarrays for the sake of access-time reduction, power reduction, and signal/noise ratio enhancement. There are two choices for memories with array division: (1) a faulty element in a subarray is replaced only by a spare element in the same subarray (intrasubarray replacement) and (2) a faulty element in a subarray may be replaced by a spare element in another subarray (intersubarray replacement). The former has smaller access penalty, while the latter realizes higher replacement efficiency. It is also possible that a subarray is replaced by a spare subarray. The devices for memorizing faulty addresses and test for finding out an effective replacement are also important issues for redundancy.

The fault distribution models are presented in Sect. 2.2. The yield improvement analysis using the models is described in Sect. 2.3. Section 2.4 describes the circuit techniques for realizing the replacement. The intrasubarray replacement, intersubarray replacement, and subarray replacement are described in Sects. 2.5, 2.6,

M. Horiguchi and K. Itoh, *Nanoscale Memory Repair*, Integrated Circuits and Systems 1,
DOI 10.1007/978-1-4419-7958-2_2, © Springer Science+Business Media, LLC 2011

and 2.7, respectively. The programmable devices for storing faulty addresses are described in Sect. 2.8. Finally, testing techniques for redundancy are explained in Sect. 2.9.

2.2 Models of Fault Distribution

2.2.1 Poisson Distribution Model

Let us consider a memory chip with N "elements" (Fig. 2.1). Here, an element may be a memory cell, a row of memory cells, a column of memory cells, a subarray, and so on. If faults are randomly distributed in the chip, the probability of an element being faulty, p, is independent of the probability of other elements being faulty or nonfaulty. Therefore, the probability that k elements are faulty and $(N - K)$ elements are not faulty is expressed as the product of their probabilities, $p^K(1 - p)^{N-K}$. Since the number of cases of selecting K faulty elements out of N elements is expressed by

$$\binom{N}{K} = {}_NC_K = \frac{N!}{(N-K)!K!} = \frac{N(N-1)\cdots(N-K+1)}{K!}, \tag{2.1}$$

the probability of existing K faulty elements in the chip is expressed as

$$P(K) = \binom{N}{K} p^K (1-p)^{N-K}. \tag{2.2}$$

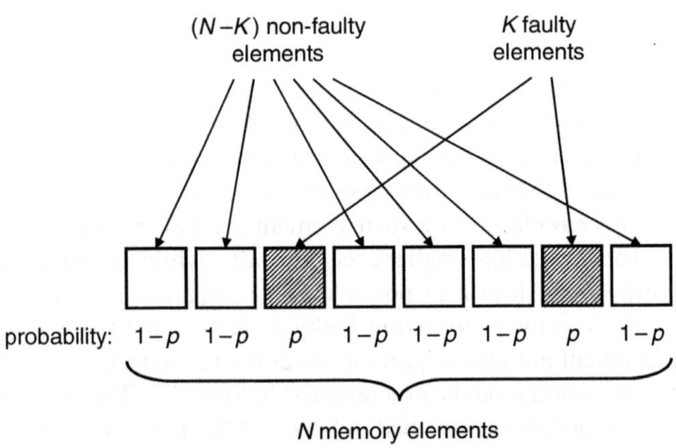

Fig. 2.1 Probability of existing K faulty elements out of N elements when faulty probability of an element is p

This is called binomial distribution and the coefficient $\binom{N}{K}$ is called binomial coefficient. Usually, N is very large and p is very small. When $N \to \infty$, keeping $\lambda = Np$ constant, (2.2) becomes

$$\begin{aligned} P(K) &= \frac{N(N-1)\cdots(N-K+1)}{K!} \cdot p^K \cdot (1-p)^{N-K} \\ &= 1 \cdot \left(1 - \frac{1}{N}\right) \cdots \left(1 - \frac{K+1}{N}\right) \cdot \frac{\lambda^K}{K!} \cdot \left(1 - \frac{\lambda}{N}\right)^N \left(1 - \frac{\lambda}{N}\right)^{-K} \\ &\to \frac{\lambda^K}{K!}\left(1 - \frac{\lambda}{N}\right)^N = \frac{\lambda^K \exp(-\lambda)}{K!} \quad (N \to \infty). \end{aligned} \tag{2.3}$$

This is called *Poisson distribution*. Figure 2.2 shows examples of the distribution. The probability $P(K)$ monotonously decreases with K for $\lambda < 1$, and has a peak around $K \sim \lambda$ for $\lambda > 1$. Poisson distribution is characterized by only one parameter λ. The average $\bar{K}$ and standard deviation $\sigma(K)$ of the number of faulty elements are expressed as

$$\bar{K} = \sum_{K=0}^{\infty} KP(K) = \exp(-\lambda) \sum_{K=1}^{\infty} \frac{\lambda^K}{(K-1)!} = \lambda, \tag{2.4}$$

$$\begin{aligned} \sigma(K) &= \sqrt{\sum_{K=0}^{\infty} K^2 P(K) - \bar{K}^2} = \sqrt{\exp(-\lambda) \sum_{K=1}^{\infty} \frac{K\lambda^K}{(K-1)!} - \lambda^2} \\ &= \sqrt{\exp(-\lambda) \left\{ \sum_{K=2}^{\infty} \frac{\lambda^K}{(K-2)!} + \sum_{K=1}^{\infty} \frac{\lambda^K}{(K-1)!} \right\} - \lambda^2} \\ &= \sqrt{\exp(-\lambda)(\lambda^2 \exp\lambda + \lambda \exp\lambda) - \lambda^2} = \sqrt{\lambda}. \end{aligned} \tag{2.5}$$

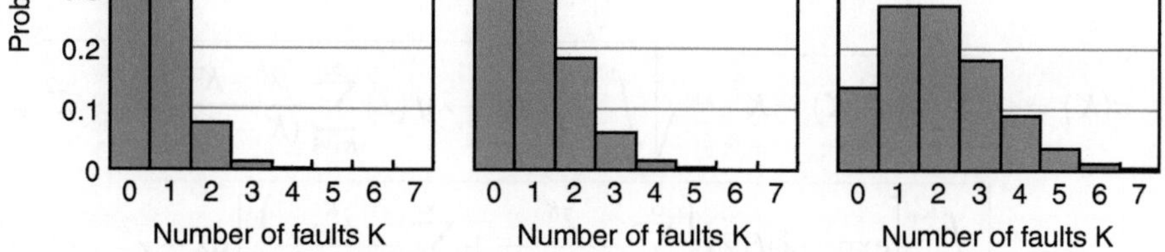

Fig. 2.2 Examples of Poisson distribution: (**a**) $\lambda = 0.5$, (**b**) $\lambda = 1.0$, and (**c**) $\lambda = 2.0$

Thus, the parameter λ is equal to the average number of faults and is expressed as:

$$\lambda = AD, \tag{2.6}$$

where A is the chip area and D is the fault density. The probability of a chip having no faulty elements (raw yield, i.e., yield without redundancy) is expressed as

$$P(0) = \exp(-\lambda) = \exp(-AD). \tag{2.7}$$

Poisson distribution model is often used for yield analysis because of its mathematical simplicity [1–5]. It is useful for rough yield estimation or the comparison of redundancy techniques. More precise yield estimation, however, requires a model that takes "fault clustering" into account described below.

2.2.2 *Negative-Binomial Distribution Model*

It has been reported that actual faults are not randomly distributed but clustered and that the number of faulty elements does not match the Poisson distribution model [6, 7]. In this case, the parameter λ is no longer constant but does distribute. Compound Poisson distribution model

$$P(K) = \int_0^\infty \frac{\lambda^K \exp(-\lambda)}{K!} \cdot f(\lambda)\mathrm{d}\lambda \tag{2.8}$$

was proposed [6] as a distribution model for nonconstant λ. The first factor in the integral is Poisson distribution and the second factor $f(\lambda)$ is a function called "compounder" representing the distribution of λ. The average $\bar{K}$ and standard deviation $\sigma(K)$ of the number of faulty elements are given by the following equations:

$$\begin{aligned}\bar{K} &= \sum_{K=0}^{\infty} KP(K) = \int_0^\infty \left\{\exp(-\lambda)f(\lambda)\sum_{K=1}^{\infty}\frac{\lambda^K}{(K-1)!}\right\}\mathrm{d}\lambda \\ &= \int_0^\infty \exp(-\lambda)f(\lambda)\cdot\lambda\exp(\lambda)\mathrm{d}\lambda = \int_0^\infty f(\lambda)\cdot\lambda\,\mathrm{d}\lambda = \bar{\lambda},\end{aligned} \tag{2.9}$$

$$\begin{aligned}\sigma(K) &= \sqrt{\sum_{K=0}^{\infty} K^2P(K) - \bar{K}^2} = \sqrt{\int_0^\infty \left\{\exp(-\lambda)f(\lambda)\sum_{K=1}^{\infty}\frac{\lambda^K\cdot K}{(K-1)!}\right\}\mathrm{d}\lambda - \bar{\lambda}^2} \\ &= \sqrt{\int_0^\infty \left[\exp(-\lambda)f(\lambda)\left\{\sum_{K=2}^{\infty}\frac{\lambda^K}{(K-2)!} + \sum_{K=1}^{\infty}\frac{\lambda^K}{(K-1)!}\right\}\right]\mathrm{d}\lambda - \bar{\lambda}^2}\end{aligned}$$

$$
\begin{aligned}
&= \sqrt{\int_0^{\infty} \{\exp(-\lambda) f(\lambda)(\lambda^2 \exp\lambda + \lambda \exp\lambda)\} \mathrm{d}\lambda - \bar{\lambda}^2} \\
&= \sqrt{\int_0^{\infty} f(\lambda)\cdot\lambda\,\mathrm{d}\lambda + \int_0^{\infty} f(\lambda)\cdot\lambda^2\,\mathrm{d}\lambda - \bar{\lambda}^2} \\
&= \sqrt{\bar{\lambda} + \{\sigma(\lambda)\}^2}.
\end{aligned}
\tag{2.10}
$$

The candidates for $f(\lambda)$ include uniform distribution and triangular distribution. However, Gamma distribution

$$
f(\lambda) = \frac{\lambda^{\alpha-1}\exp(-\lambda/\beta)}{\Gamma(\alpha)\beta^{\alpha}} \tag{2.11}
$$

has been shown the most suitable for actual fault distribution[1] [6, 8]. The meanings of the parameters α and β are as follows: α corresponds to fault clustering (smaller α means stronger clustering), and the product $\alpha\beta$ is equal to the average of λ, λ_0. The standard deviation of λ is equal to $\beta\sqrt{\alpha}$. Figure 2.3 shows examples of (2.11) for various parameters maintaining $\lambda_0 = \alpha\beta = 1.0$. When $\alpha \to \infty$, the distribution becomes the delta function, corresponding to no λ distribution.

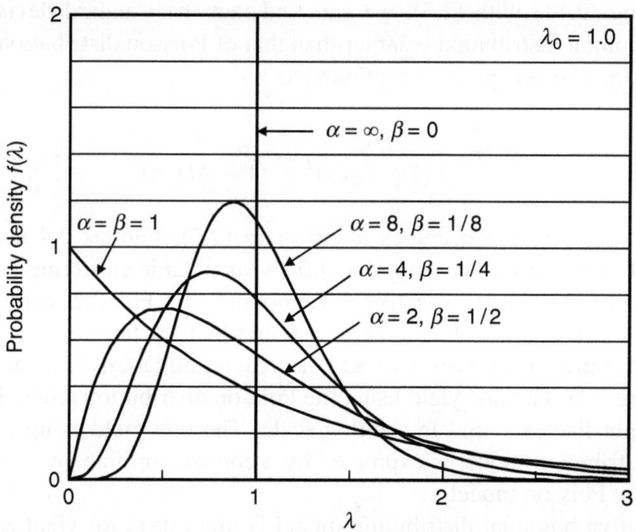

Fig. 2.3 Probability density function of gamma distribution as a compounder (average of $\lambda = 1.0$)

[1] $\Gamma(\alpha)$ is gamma function defined as $\Gamma(\alpha) = \int_0^{\infty} t^{\alpha-1}\exp(-t)\mathrm{d}t$. $\Gamma(\alpha) = (\alpha - 1)!$ for integer α.

Substituting (2.11) and $\beta = \lambda_0/\alpha$ into (2.8) results in

$$
\begin{aligned}
P(K) &= \int_0^\infty \frac{\lambda^K \exp(-\lambda)}{K!} \cdot \frac{\lambda^{\alpha-1} \exp(-\lambda/\beta)}{\Gamma(\alpha)\beta^\alpha} \mathrm{d}\lambda \\
&= \frac{1}{K!\Gamma(\alpha)\beta^\alpha} \int_0^\infty \lambda^{K+\alpha-1} \exp\left\{-\left(1+\frac{1}{\beta}\right)\lambda\right\} \mathrm{d}\lambda \\
&= \frac{1}{K!\Gamma(\alpha)\beta^\alpha} \left(\frac{\beta}{\beta+1}\right)^{K+\alpha} \int_0^\infty t^{k+\alpha-1} \exp(-t)\mathrm{d}t \\
&= \frac{\Gamma(K+\alpha)\beta^K}{K!\Gamma(\alpha)(\beta+1)^{K+\alpha}} \\
&= \frac{\alpha(\alpha+1)\cdots(\alpha+K-1)(\lambda_0/\alpha)^K}{K!(1+\lambda_0/\alpha)^{K+\alpha}}.
\end{aligned}
\tag{2.12}
$$

This is called *negative binomial distribution* [9]. The average and standard deviation of the number of faulty elements are calculated from (2.9) and (2.10):

$$\bar{K} = \bar{\lambda} = \lambda_0, \tag{2.13}$$

$$\sigma(K) = \sqrt{\bar{\lambda} + \{\sigma(\lambda)\}^2} = \sqrt{\lambda_0 + \beta^2\alpha} = \sqrt{\lambda_0(1 + \lambda_0/\alpha)}. \tag{2.14}$$

Comparing (2.14) with (2.5), we can find that the standard deviation of the negative-binomial distribution is larger than that of Poisson distribution by a factor of $\sqrt{1 + \lambda_0/\alpha}$. The raw yield is expressed as

$$P(0) = \frac{1}{(1 + \lambda_0/\alpha)^\alpha} = \frac{1}{(1 + AD/\alpha)^\alpha}. \tag{2.15}$$

When $\alpha \to \infty$, (2.15) becomes identical to (2.7). Figures 2.4 and 2.5 show examples of the distribution with $\alpha = 4.0$ (weaker fault clustering) and $\alpha = 1.0$ (stronger fault clustering), respectively. Compared with Fig. 2.2 (corresponding to the case $\alpha = \infty$), the probability for $K = 0$ and that for large K increase and the probability for medium K decreases as α decreases. Equations (2.7) and (2.15) are plotted in Fig. 2.6. The raw yield using the Poisson distribution model is expressed by the straight line ($\alpha = \infty$) in semilog scale. The raw yield using the negative-binomial distribution model is expressed by a concave-up line and is greater than that using the Poisson model.

The negative-binomial distribution model is often used for yield estimation of memory LSIs [10–12] because it gives good agreement with actual fault distribution. In order to use this model, however, we must determine two parameters λ_0 (average number of faults) and α (fault clustering factor) from experimental data. In addition, it should be noted that the parameter α may depend on the kind of the memory element.

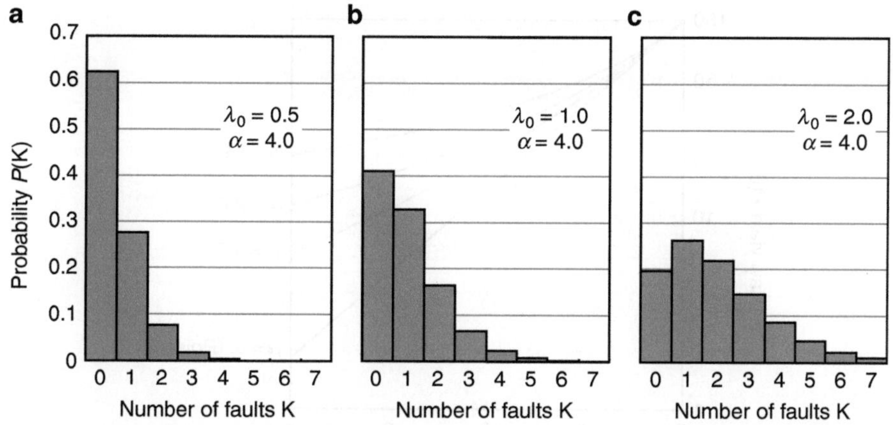

Fig. 2.4 Examples of negative binomial distribution with $\alpha = 4.0$: (**a**) $\lambda_0 = 0.5$, (**b**) $\lambda_0 = 1.0$, and (**c**) $\lambda_0 = 2.0$

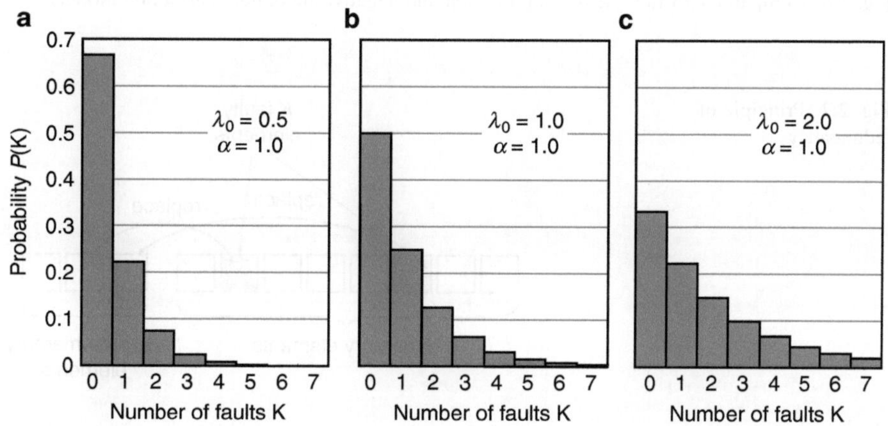

Fig. 2.5 Examples of negative binomial distribution with $\alpha = 1.0$: (**a**) $\lambda_0 = 0.5$, (**b**) $\lambda_0 = 1.0$, and (**c**) $\lambda_0 = 2.0$

2.3 Yield Improvement Through Redundancy

In this section, yield improvement through redundancy is analyzed using the models described above. We assume the followings for simplicity:

1. Faults on spare elements are neglected.
2. Fatal faults are neglected. A fatal fault is defined as a fault that makes the entire chip no good. For example, a defect on peripheral circuit of a memory LSI may cause a fatal fault.

Without redundancy, the yield Y_0 is equal to $P(0)$ as shown in Fig. 2.6 because only chips without faulty elements are accepted. If R spare elements

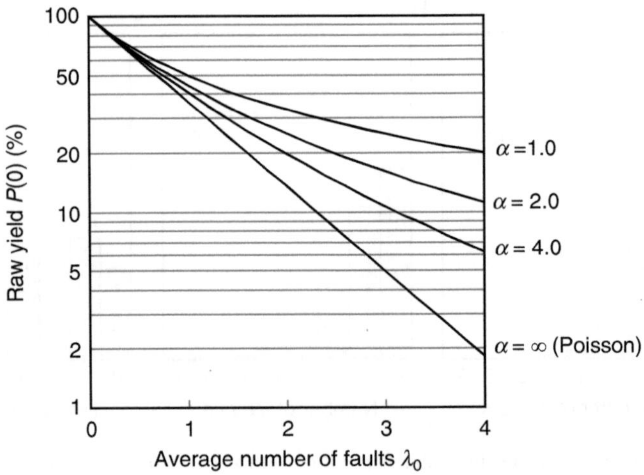

Fig. 2.6 Comparison of raw yield using Poisson and negative-binomial distribution models

Fig. 2.7 Principle of redundancy

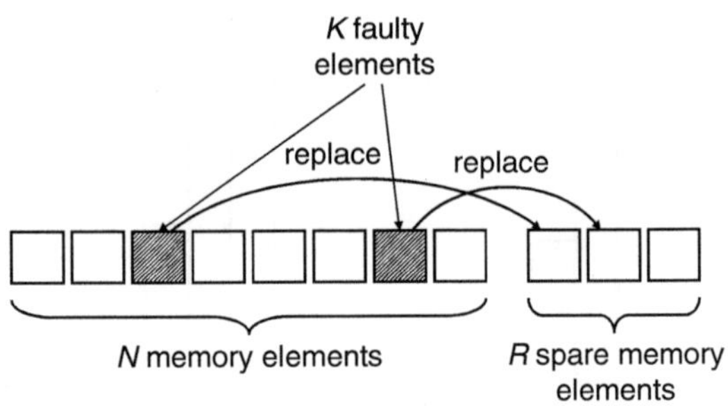

are added in the chip, chips with K faulty elements ($K \leq R$) become acceptable by replacing the faulty elements with spares as shown in Fig. 2.7. Therefore, the yield becomes

$$Y = \sum_{K=0}^{R} P(K). \tag{2.16}$$

Figures 2.8 and 2.9 show the calculated yield using the Poisson distribution and the negative-binomial distribution models, respectively. The raw yield Y_0 ($R = 0$) is lower but the yield improvement is larger with Poisson distribution model than with negative-binomial model. This is apparent in Figs. 2.10 and 2.11, where the relationships between yields with and without redundancy are plotted. Thus, it should be noted that using Poisson distribution model tends to underestimate Y_0 and overestimate the yield improvement.

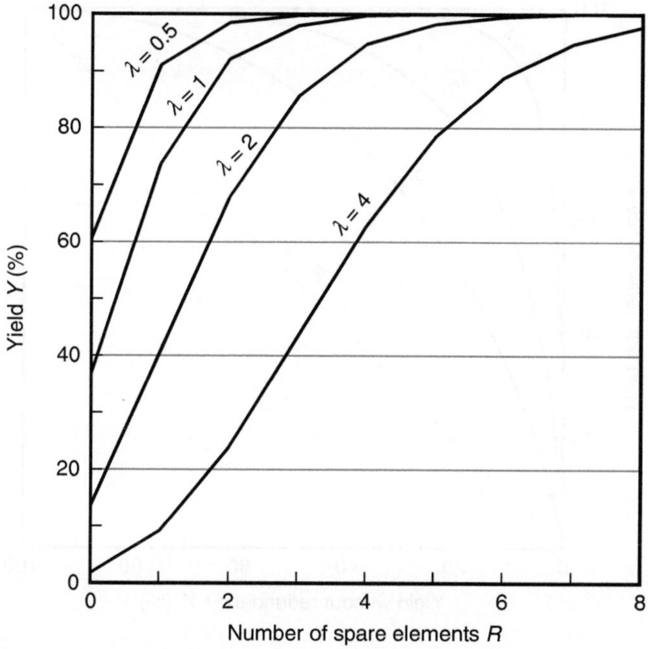

Fig. 2.8 Yield improvement through redundancy using Poisson distribution model

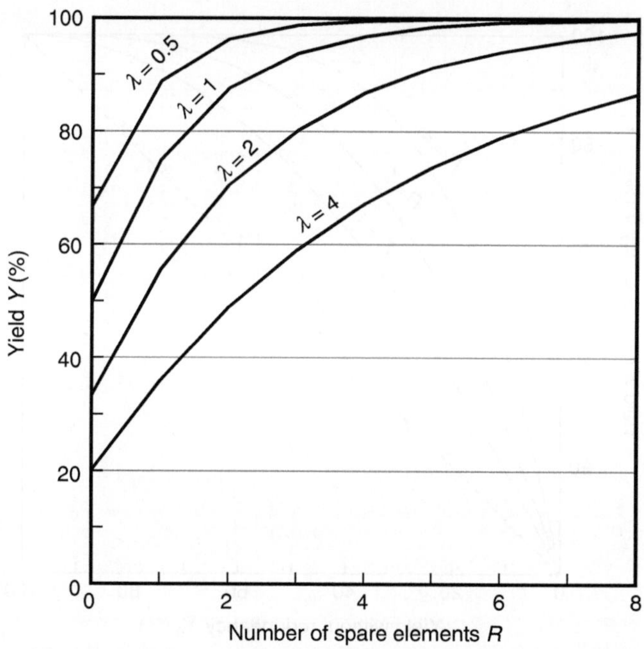

Fig. 2.9 Yield improvement through redundancy using negative-binomial distribution model ($\alpha = 1.0$)

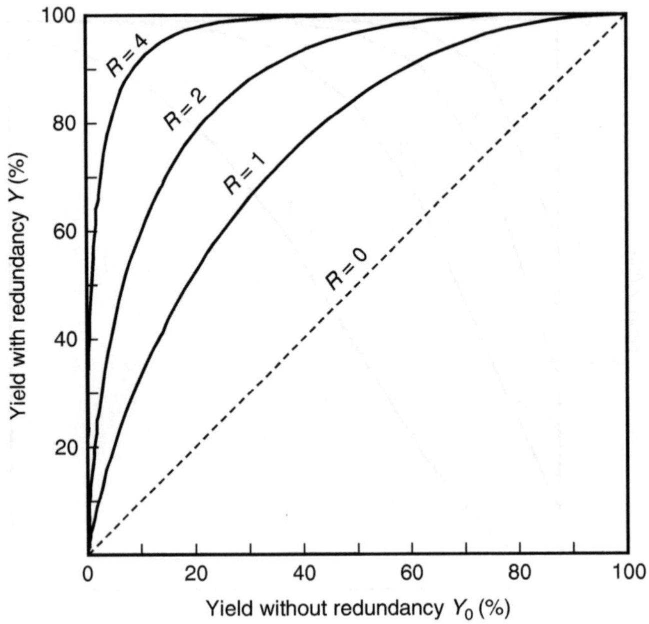

Fig. 2.10 Yield improvement through redundancy using Poisson distribution model: number of spare elements as a parameter

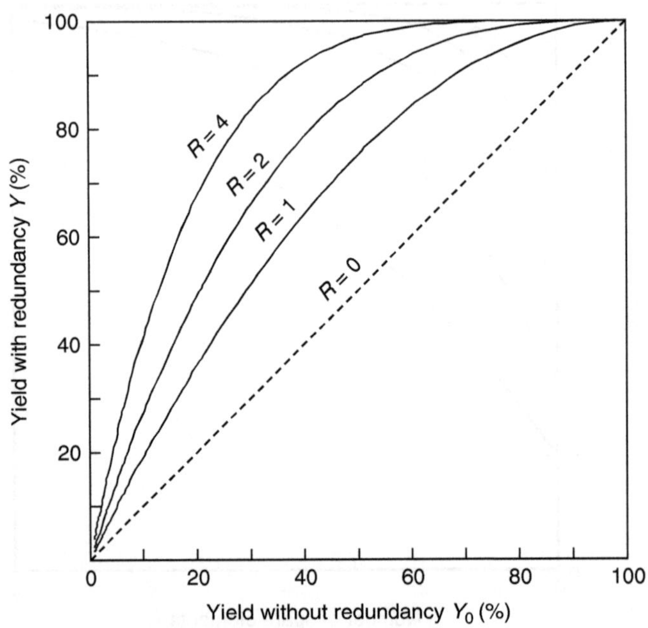

Fig. 2.11 Yield improvement through redundancy using negative-binomial distribution model ($\alpha = 1.0$): number of spare elements as a parameter

2.4 Replacement Schemes

2.4.1 *Principle of Replacement*

Since redundancy techniques repair memory LSIs by replacing faulty normal elements by spare elements and by hiding the faulty elements from users, the following steps are required.

1. Replacement information (which normal element is replaced by which spare element) is stored in on-chip programmable devices in advance.
2. When accessed, it is judged whether the demanded address is faulty ("hit") or not ("miss") using the information stored above.
3. In case of miss, the normal element is activated and no spare elements are activated.
4. In case of hit (a) the normal element is inhibited from being activated and (b) the spare element that replaces the normal element is instead activated.

Step 1 requires programmable devices. In addition, they must be nonvolatile to retain the programmed replacement information during power-off. The devices used for this purpose include fuses, antifuses, and nonvolatile memory cells as is described in Sect. 2.8. There are two schemes for storing and reading the replacement information in the storage (steps 1 and 2): decoder programming and address comparison. The former utilizes spare decoders, which is the same as normal decoders except that the address is programmable (Fig. 2.12). The latter utilizes comparators for selecting spare elements (Fig. 2.13). In addition, there are two schemes for disabling the faulty normal element (step 4a): direct disabling and

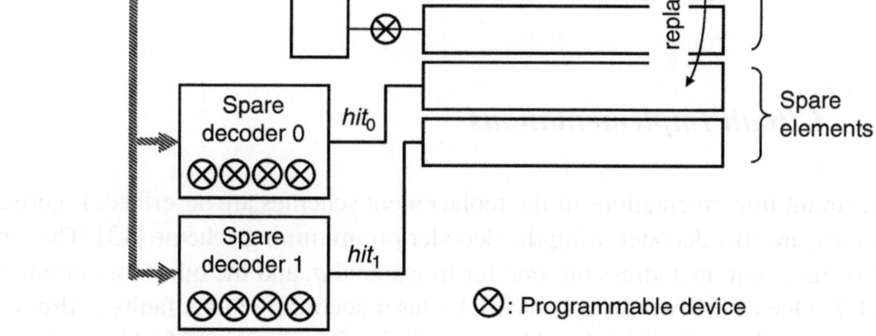

Fig. 2.12 Replacement using decoder programming and direct disabling schemes

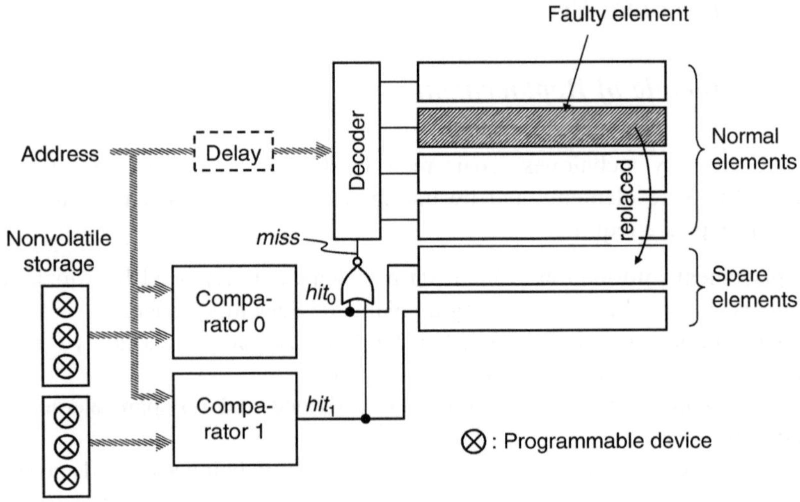

Fig. 2.13 Replacement using address comparison and indirect disabling schemes

indirect disabling. The former cuts off the signal path to the faulty element by blowing a fusible link (Fig. 2.12). The latter utilizes the spare-selection signals hit_i (Fig. 2.13). The activation of normal element is inhibited if one of the hit_i is asserted, and is allowed if none of the hit_i is asserted. Therefore, there are four (2×2) possible combinations. Figure 2.12 shows a replacing scheme using decoder programming and direct disabling schemes, while Fig. 2.13 shows a scheme using address comparison and indirect disabling schemes. On the other hand, there is a quite different replacing scheme, shifting scheme, as shown in Fig. 2.14. The switches inserted between the decoder and the memory array are controlled by the programmable storage, in which the address of a faulty element is programmed. In case of no faulty elements, each output of the decoder is connected with the corresponding normal element and the spare element is not connected. If there is a faulty normal element, the connections are shifted so that the outputs of the decoder are connected with the spare element and the normal elements except for the faulty one.

2.4.2 *Circuit Implementations*

Next, circuit implementations of the replacement schemes are described. Figure 2.15 shows a spare row decoder using the decoder programming scheme [13]. The circuit has two fuses per an address bit, one for true signal a_i and the other for complement signal $\bar{a}_i$. One of the two fuses is blown by laser according to the faulty address. The node N is precharged to high level by signal PC before the input of address signals. If the input address coincides with the programmed address the node N remains high.

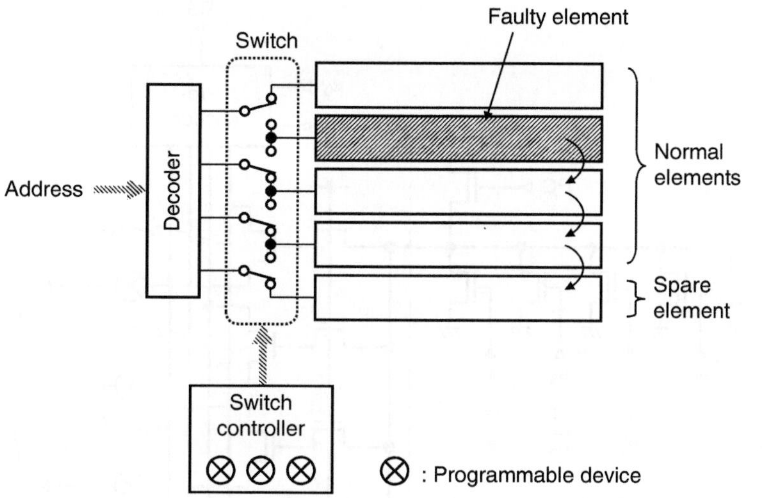

Fig. 2.14 Replacement using shifting scheme

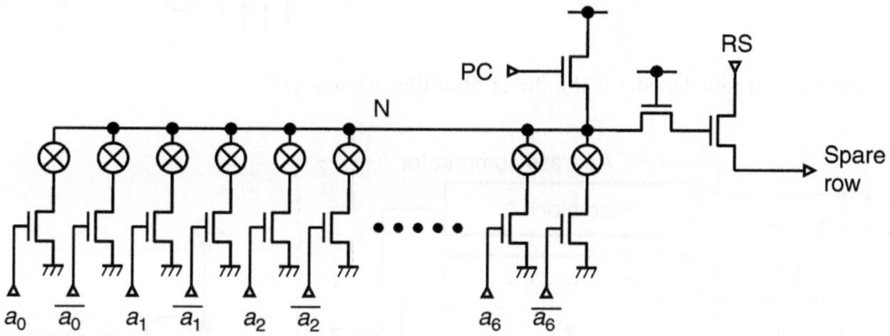

Fig. 2.15 Spare row decoder using decoder programming scheme. Reproduced from [13] with permission; © 2010 IEEE

The spare row line is activated when row selection signal RS goes high. If the input address does not coincide with the programmed address, node N goes low and the spare line is not activated. If no faulty address, all the fuses are unblown. Since node N always goes low irrespective of the input address, the spare row line is never activated. Figure 2.16 shows a normal row decoder using the direct disabling scheme [13]. The output of the NOR gate is ANDed with the predecoder output signals RS_0–RS_3 to relax the circuit pitch. A normal row is disconnected to the output of the decoder by blowing the corresponding fuse by laser. Thus, this circuit requires as many fuses as the rows and the pitch of fuses is equal to row pitch. Figure 2.17 shows a row redundancy circuit using the address comparison and indirect disabling schemes [14]. Each address comparator unit stores a faulty address and compares it with the input address to generate the spare-selection signal hit_i. If hit_i is asserted, the corresponding spare row is activated at the timing

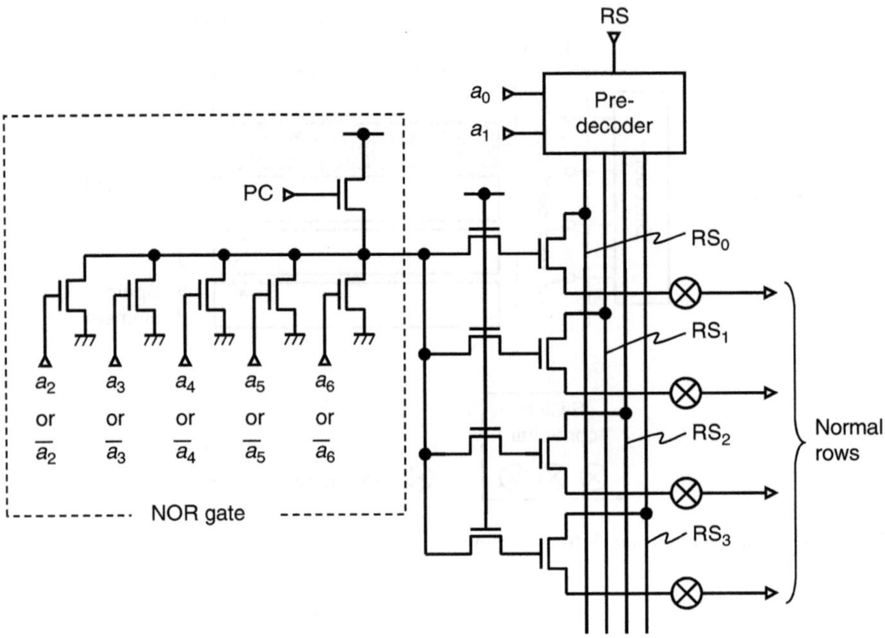

Fig. 2.16 Normal row decoder using direct disabling scheme [13]

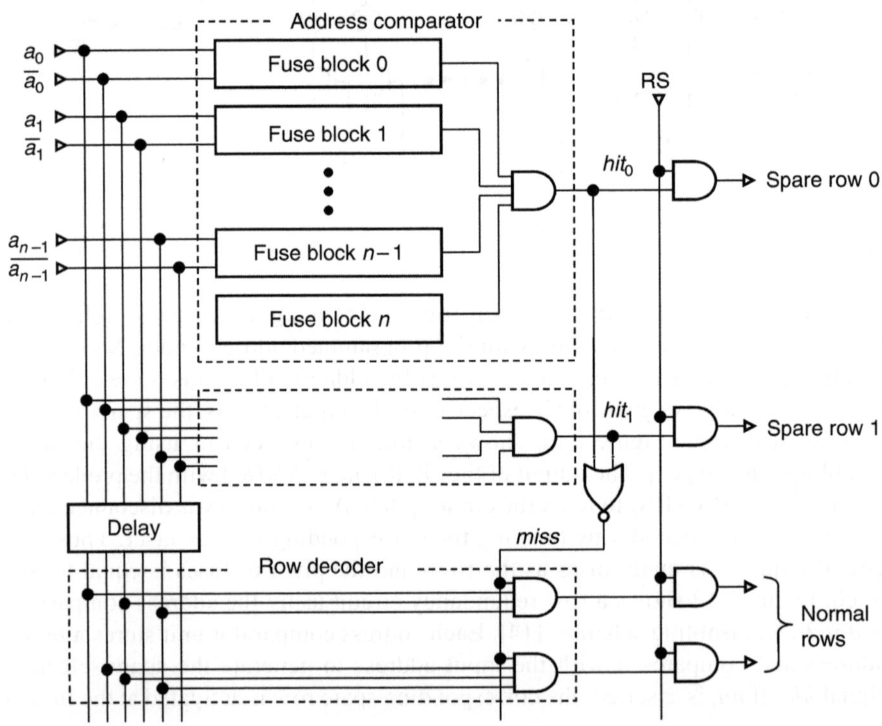

Fig. 2.17 Row replacement using address comparison and indirect disabling schemes [14]

of RS and the row decoder is disabled through the NOR gate. Note that a delay circuit is inserted at the input of the row decoder. Without this delay, the arrival of the address signals at the row decoder precedes the generation of hit_i, and a normal row may be wrongly activated. Therefore, this circuit has an access time penalty of the delay time. The detail of the fuse block included in the address comparator is shown in Fig. 2.18 [14, 15]. This circuit utilizes an electrically blown fuse. The blowing MOS transistor M_0 is controlled by programming signal $\bar{P}$ and address signal $\bar{a}_i$. If the fuse is blown, the signal $\bar{f}_i$ is pulled down to low and f_i is high. Since M_1 is on and M_2 is off, the output signal c_i is equal to a_i. On the contrary, if the fuse is unblown, c_i is equal to $\bar{a}_i$. The programming and normal operations of the fuse block are summarized in Table 2.1. Note that $c_i = 1$ if the address during programming is equal to the address during normal operation and otherwise $c_i = 0$. All the outputs of the fuse blocks are ANDed to generate the hit signal and to activate the corresponding spare row line. An extra fuse block is added to indicate whether the address comparator is valid or not. Without blowing the fuse in this extra block, the comparator is invalid and the corresponding spare row line is never activated.

The number of fuses required for each scheme is as follows. The decoder programming scheme require $2nR$, while the address comparison requires $(n + 1)$ R (including the fuses for extra blocks), where $n = \log_2 N$ is the number of address bits, N is the number of normal lines, and R is the number of spare lines. The direct disabling scheme requires 2^n because the number of normal lines is $N = 2^n$ while the indirect disabling scheme requires none. Therefore, the combination of address

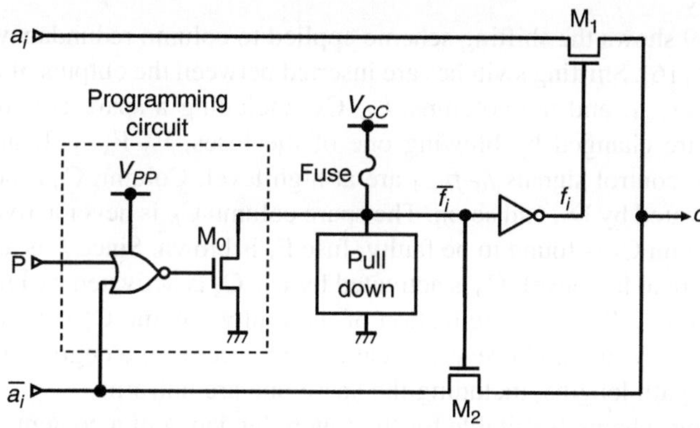

Fig. 2.18 Detail of fuse block [14, 15]

Table 2.1 Truth table of fuse block

Programming ($\bar{P} = 0$)					Normal operation		
a_i	$\bar{a}_i$	Fuse	f_i	$\bar{f}_i$	a_i	$\bar{a}_i$	c_i
0	1	Unblown	0	1	0	1	1
					1	0	0
1	0	Blown	1	0	0	1	0
					1	0	1

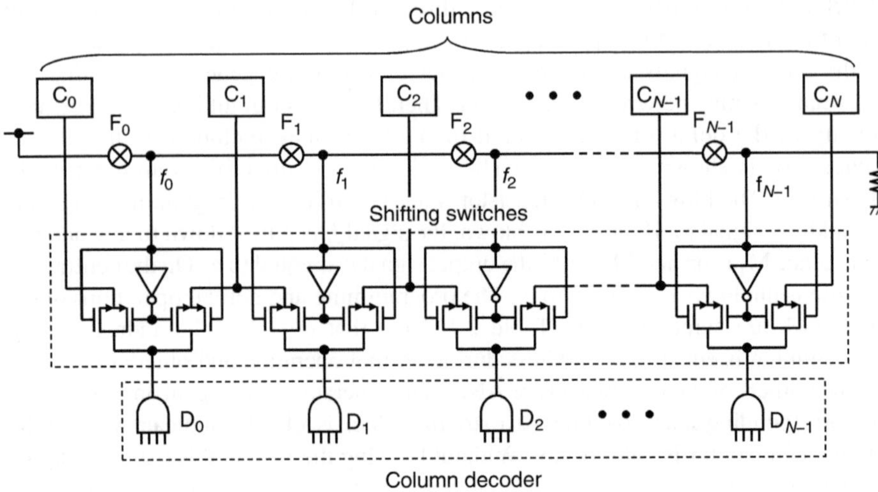

Fig. 2.19 Column replacement using shifting scheme. Reproduced from [16] with permission; © 2010 IEEE

comparison and indirect disabling schemes requires the fewest fuses. Since the size of fuses is not so scaled down according to design rules, the combination is the best choice for high-density memory LSIs with a large number of spare lines, despite the access penalty.

Figure 2.19 shows the shifting scheme applied to column redundancy of a high-speed SRAM [16]. Shifting switches are inserted between the outputs of the column decoder, D_0–D_{N-1}, and the columns, C_0–C_N, including a spare column C_N. The connections are changed by blowing one of the fuses, F_0–F_{N-1}. If no fuses are blown, all the control signals f_0–f_{N-1} are at high level. Column C_0 is activated by D_0, C_1 is activated by D_1, and so on. The spare column C_N is never activated. If, for example, column C_1 is found to be faulty, fuse F_1 is blown. Since f_0 is at high level and $f_1 - f_{N-1}$ are at low level, C_0 is activated by D_0, C_2 is activated by D_1, and C_N is activated by D_{N-1}. Thus, the activation of the faulty column C_1 is inhibited. This scheme requires N fuses. However, it features no access-time degradation because all the signal-path lengths, including the spare one are uniform.

The shifting scheme is suitable for the row redundancy of a content addressable memory (CAM) because a CAM has a priority encoder (PE). If the search data matches the data stored in two or more rows, the PE outputs the address of the row having the highest priority. The order of priority is usually ascending or descending order of row address. If a faulty row is replaced by a spare row with the ordinary replacing scheme, the order of address is not maintained. With the shifting scheme, however, the order of row address is maintained even after shifting [17, 18]. Figure 2.20 shows a CAM with row redundancy using the shifting scheme. The information of faulty address stored in the fuses is decoded and is transferred to the register FR in advance and is used for switching both wordlines and matchlines.

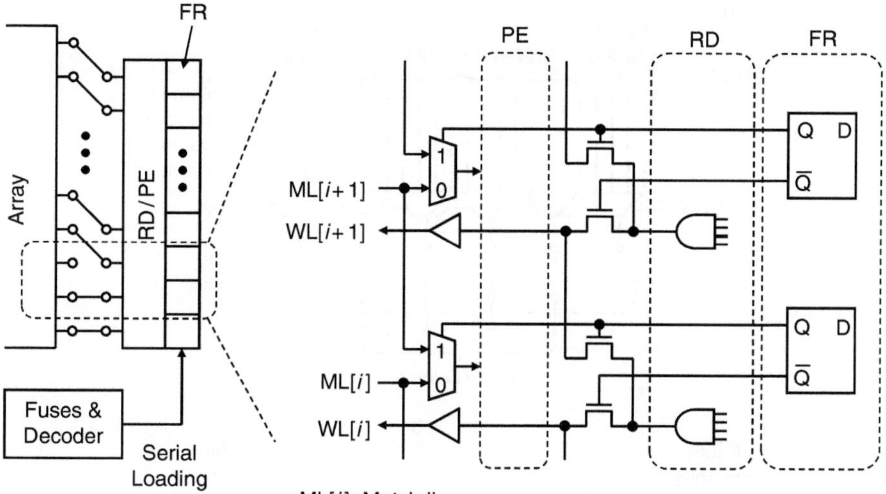

Fig. 2.20 Row replacement of a CAM using shifting scheme. Reproduced from [17] with permission; © 2010 IEEE

The shifting scheme is also suitable for wide I/O memories. Figure 2.21 shows the shifting scheme applied to a DRAM macro with 128 I/Os [19]. The macro has 128 normal blocks and two spare blocks on both sides. The redundancy scheme applied to this macro features that the connections of switches are variable according to column address (the connections of the switches in Figs. 2.19 and 2.20 are fixed after blowing fuses). A column of each block is selected by multiplexer MUX controlled by column selection lines CSL. When column 0 is selected (Fig. 2.21a), only normal blocks are connected to the I/O lines because column 0 of each block is not faulty. However, when another column is selected (Fig. 2.21b, c), the connections are changed so that the faulty columns are not connected to the I/O lines. The detail of the shifting switch is shown in Fig. 2.22. Faulty block addresses of each column are programmed in the shift point register SPR by blowing fuses. The faulty block addresses of the selected column, FB_0 and FB_1, are read out and compared with each block address generated by a wired logic circuit. The connection of each switch is determined by the comparison results as shown in the table.

Although the circuit in Fig. 2.19 requires N fuses, the circuit in Fig. 2.20 requires only $\log_2 N + 1$ because of decoding. The circuit in Fig. 2.21 requires $2(\log_2 N + 1)$ because there are two spare columns. The number of programmable elements required for replacement schemes is summarized in Table 2.2.

A disadvantage of the shifting scheme is that the number of spares is limited. Since R sets of spares require $(R + 1)$-terminal shifting switches, a large R causes the complexity of switches and the control circuits. Therefore, $R = 1$ or 2 is practical.

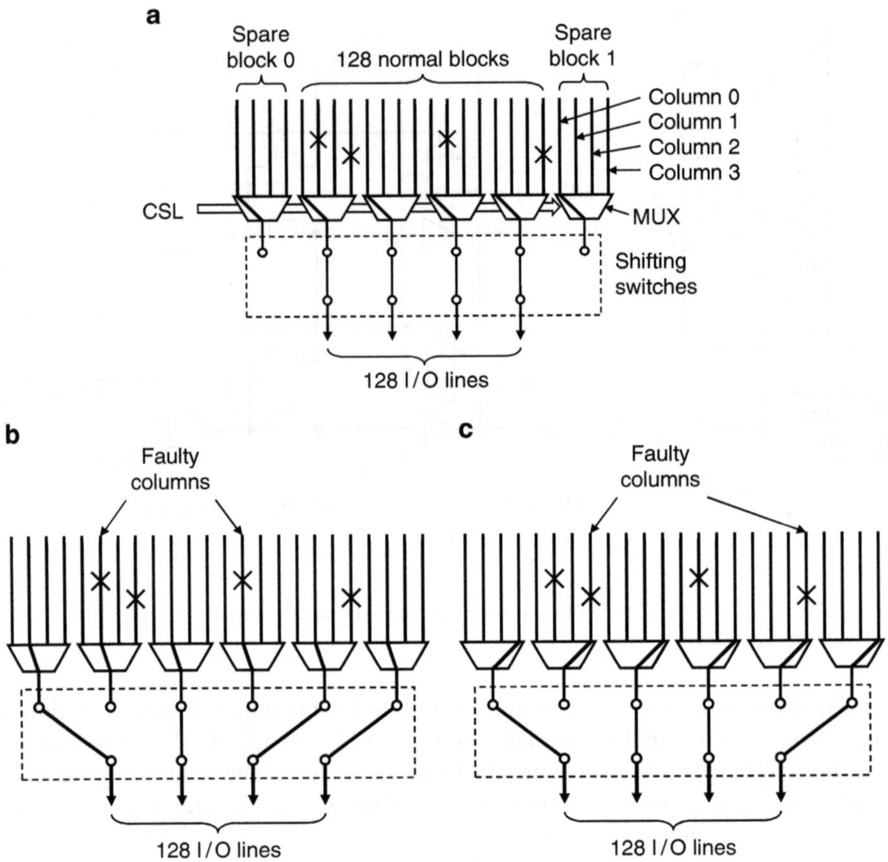

Fig. 2.21 Flexible DQ replacement of an embedded RAM using shifting scheme: (**a**) nonfaulty case, (**b**) two faulty columns are replaced, and (**c**) when another address is selected, other two faulty columns are replaced. Reproduced from [19] with permission; © 2010 IEEE

2.5 Intrasubarray Replacement

One of the problems for redundancy with the increase in memory capacity is memory-array division. Figure 2.23 shows the trend of memory-array division of DRAMs [20]. The number of subarrays doubled each generation before 64 Mbit. This is mainly due to the dataline (bitline) division shown in Fig. 2.24. The dataline D is divided into D_0–D_3 to reduce the parasitic capacitance for signal/noise ratio enhancement and charging/discharging current reduction [21–24]. The number of divisions even quadrupled each generation after the introduction of hierarchical wordline structure as shown in Figs. 2.25 and 2.26 [25–27]. Here, a wordline is divided into a plurality of sub wordlines (SWLs), each of which is activated by the AND of a main wordline (MWL) and a block selection line BS.

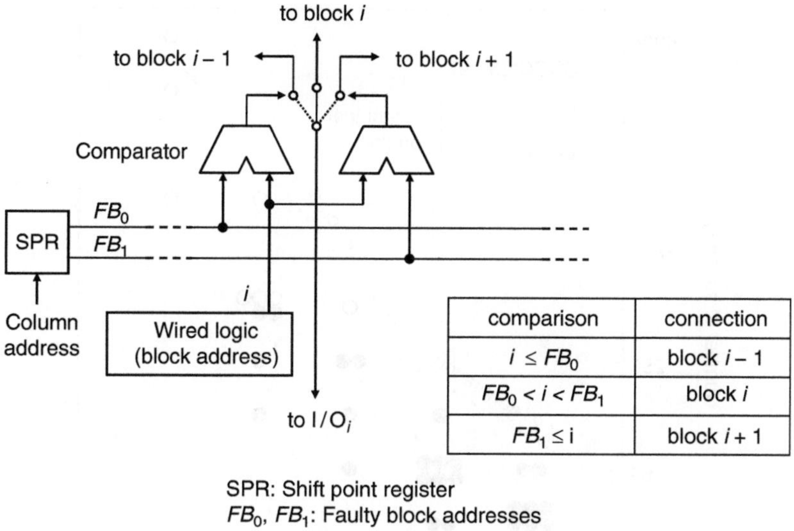

Fig. 2.22 Detail of shifting switch and its control circuit [19]

Table 2.2 Number of programmable elements required for replacement

Replacement scheme	Number of programmable elements	Figure
Decoder programming + direct disabling	$2R \log_2 N + N$	2.15, 2.16
Decoder programming + indirect disabling	$2R \log_2 N$	–
Address comparison + direct disabling	$R(\log_2 N + 1) + N$	–
Address comparison + indirect disabling	$R(\log_2 N + 1)$	2.17
Shifting (without decoding)	$N\ (R = 1)$	2.19
Shifting (with decoding)	$R(\log_2 N + 1)$	2.20, 2.21

These memory array divisions degrade yield because the boundaries between subarrays work as barriers to the faulty-element replacement and reduce the replacement flexibility.

There are two approaches to cope with this problem. One is to enhance the replacement flexibility under the restriction of intrasubarray replacement, that is, a faulty element is replaced only by a spare element in the same subarray. The other is to allow intersubarray replacement, that is, a faulty element can be replaced by a spare element in another subarray. The former is explained in this section and the latter is described in the next section.

Here, let us define two terms used in the following discussion. *Replacement unit* is defined as a set of memory cells replaced simultaneously (by programming one set of programmable devices). The replacement unit is assumed as a row/column so far. However, adjacent two rows/columns are often replaced simultaneously in actual memory design. This is realized by neglecting the least significant bit of row/column address. In this case the replacement unit is two rows/columns. In addition, in the case of the simultaneous replacement described below, all the rows/columns in all the subarrays are replaced simultaneously. In this case the

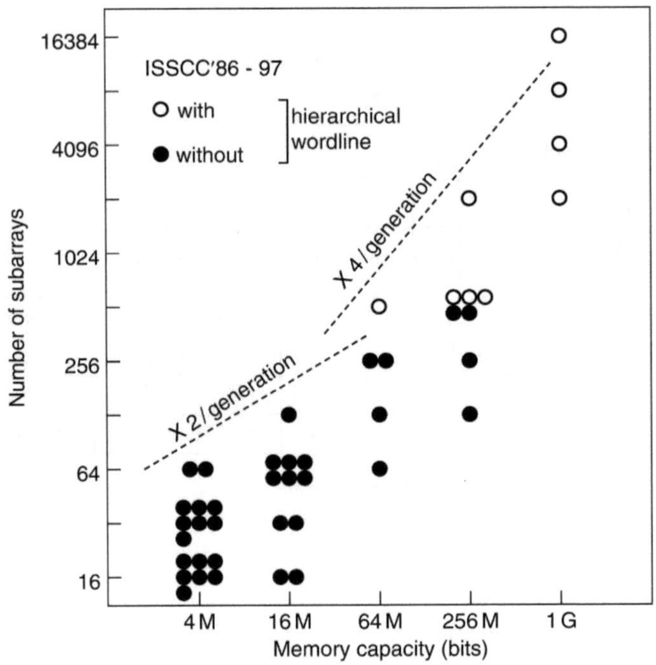

Fig. 2.23 Trend of DRAM memory-array division. Reproduced from [20] with permission; © 2010 IEEE

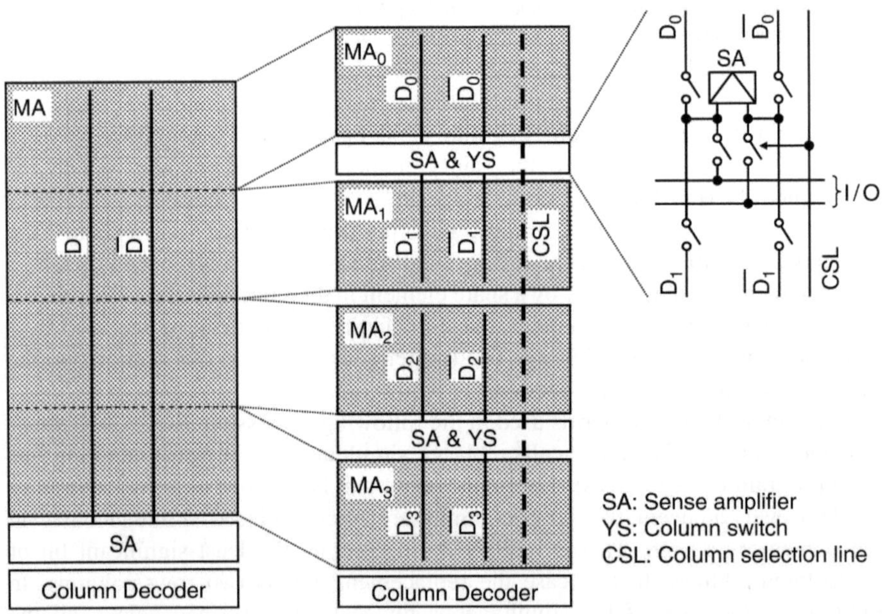

Fig. 2.24 Memory-array division using multidivided dataline (bitline) structure [21, 23, 24]

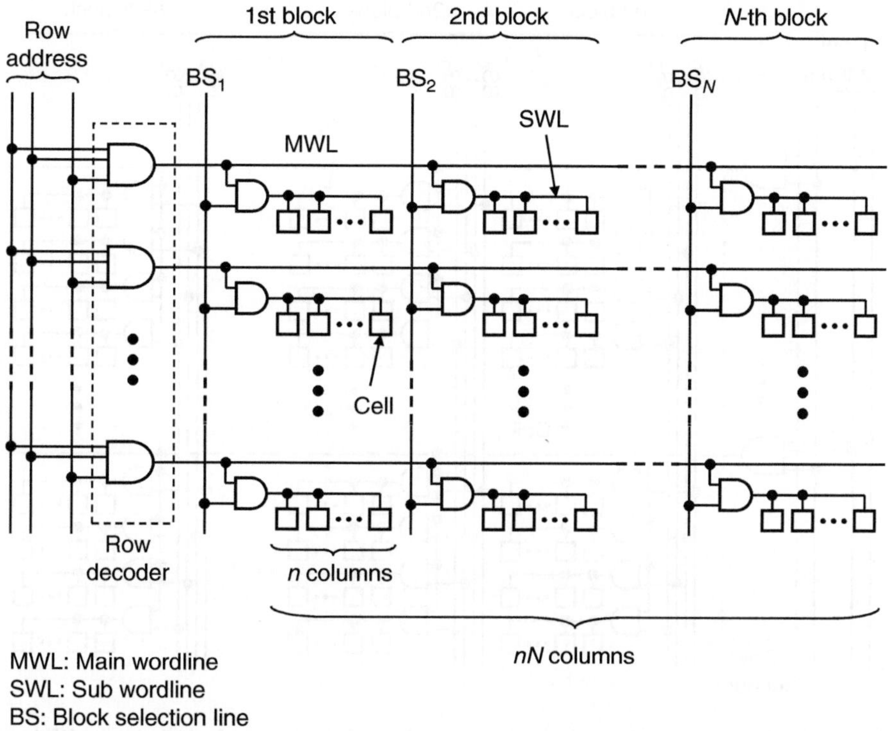

Fig. 2.25 Memory-array division using hierarchical wordline structure for high-speed SRAMs [25]

replacement unit is *M* rows/columns, where *M* is the number of subarrays. *Replacement region* is defined as an area in which any normal row/column can be replaced by any spare row/column. The replacement region is a subarray in the case of intrasubarray replacement, while it is plural subarrays in the case of intersubarray replacement.

2.5.1 Simultaneous and Individual Replacement

Figure 2.27 shows the row redundancy technique applied to a memory without array division. The memory has L (here, $L = 4$) spare wordlines SW_0–SW_3 and as many address comparators AC_0–AC_3. Faulty word addresses are programmed in the address comparators and are compared with the input address. Thus, at most L faulty normal wordlines can be repaired. In this example, faulty normal wordlines W_0–W_3 are replaced by spare wordlines SW_0–SW_3, respectively, as shown by the arrows in the figure. Now let us consider dividing the memory array into subarrays. Two approaches within the restriction of intrasubarray replacement are shown in Figs. 2.28 and 2.29. Here the memory array MA in Fig. 2.27 is divided into four

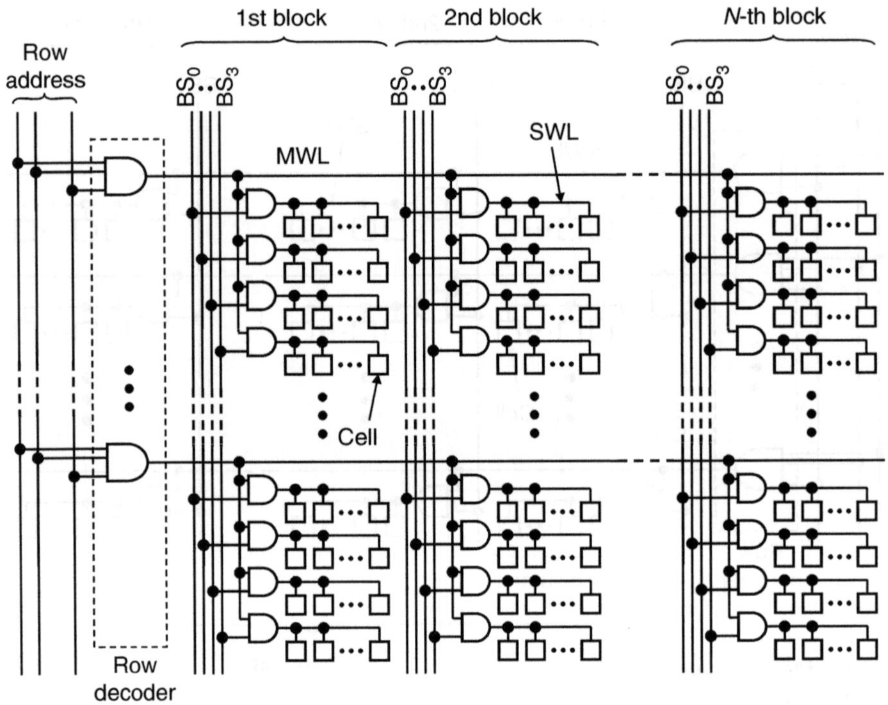

Fig. 2.26 Memory-array division using hierarchical wordline structure for high-density DRAMs [26, 27]

subarrays MA_0–MA_3, only one of which is activated. Among m row address bits, the upper two bits a_{m-2} and a_{m-1} (intersubarray address signals) are used for selecting a subarray and the lower $(n - 2)$ bits a_0–a_{m-3} (intrasubarray address signals) are used for selecting a wordline in the subarray.

In the simultaneous replacement (Fig. 2.28), the number of address comparators is equal to L, the number of spare wordlines in each subarray. The total number of spare wordlines is therefore LM, where $M(=4)$ is the number of subarrays. Each address comparator compares only the intrasubarray address signals, and the output is commonly supplied to every subarray. The intersubarray address signals, in turn, select one of the four spare wordlines. As many faulty wordlines can be repaired as are shown in Fig. 2.27, if L is the same. In this approach, four normal lines are replaced simultaneously by spare lines as shown by the arrows in Fig. 2.28. That is, to replace one faulty line, three other normal lines with the same intrasubarray address are also replaced even if they are not faulty. The replacement unit is therefore four wordlines. This causes the following problems. First, the usage efficiency of spare lines is lower (25% in this case) and the number of spare lines should be larger, resulting in chip-area increase. Second, the probability of unsuccessful repair due to faults on the spare lines that replaced normal lines is higher, resulting in yield degradation.

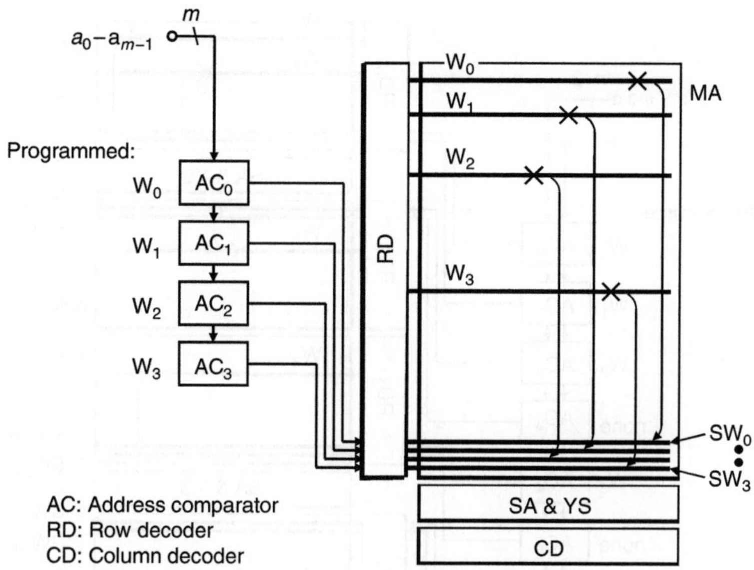

Fig. 2.27 Row redundancy applied to a memory without array division

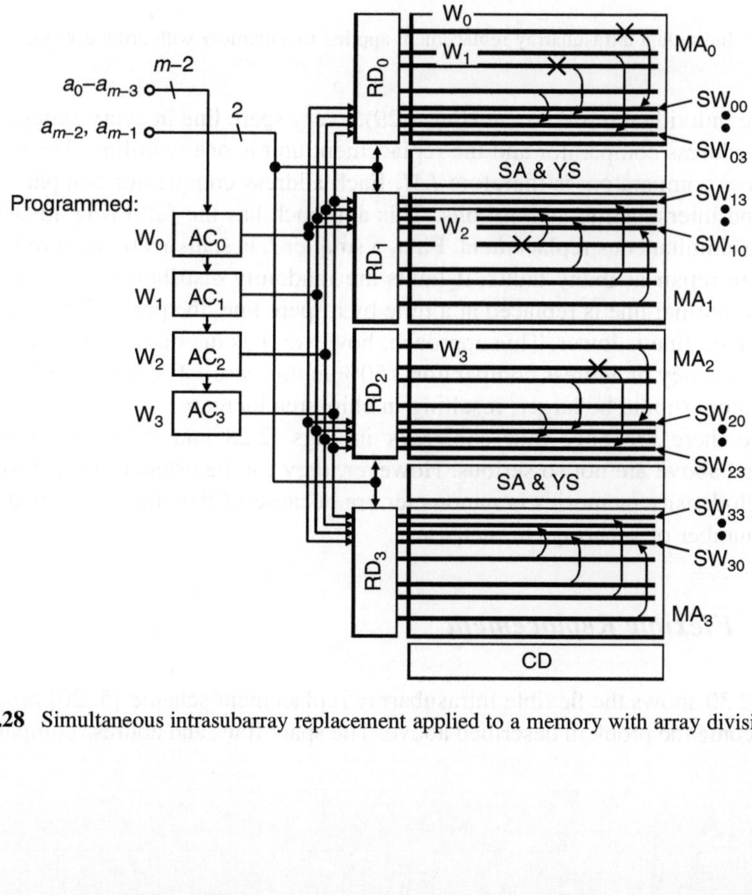

Fig. 2.28 Simultaneous intrasubarray replacement applied to a memory with array division

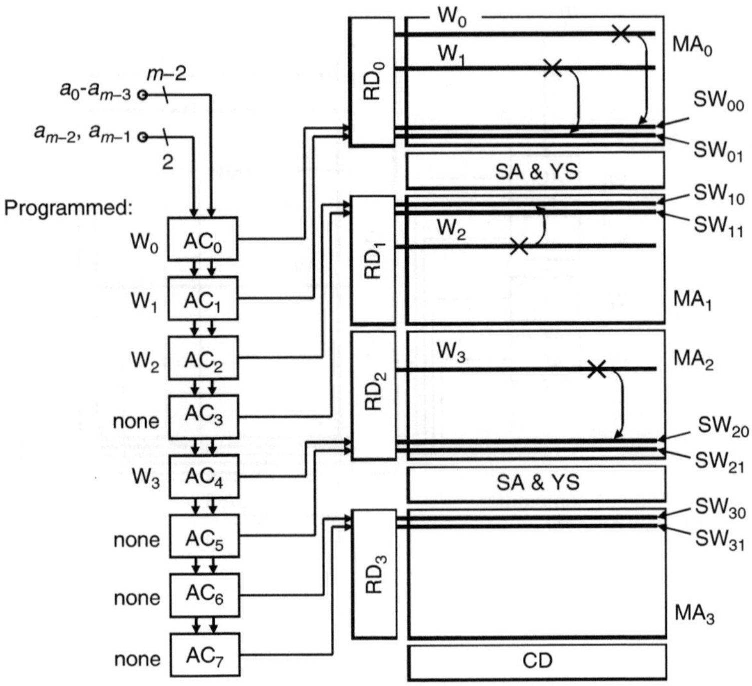

Fig. 2.29 Individual intrasubarray replacement applied to a memory with array division

In the individual replacement (Fig. 2.29), every spare line in every subarray has its own address comparator and the replacement unit is one wordline. The number of address comparators is therefore LM. Each address comparator compares both intra- and intersubarray address bits. This approach has the following advantages over the simultaneous replacement. First, a smaller L is statistically required (here, $L = 2$) to repair as many faults, if faults are randomly distributed. Second, since only one normal line is replaced at a time by a spare line, the probability of a fault on the spare line is lower. This approach, however, has the disadvantage of lower usage efficiency of address comparators (50% in this case). The number of address comparators should be larger, resulting in chip-area increase.

Since there are only four subarrays in Figs. 2.28 and 2.29, the problems described above are not so serious. However, they can be critical in the design of ultrahigh-density memories in nanometer era because of the aforementioned trend of the number of subarrays to increase.

2.5.2 *Flexible Replacement*

Figure 2.30 shows the flexible intrasubarray replacement scheme [5, 20] proposed to overcome the problem described above. The spare lines and address comparators

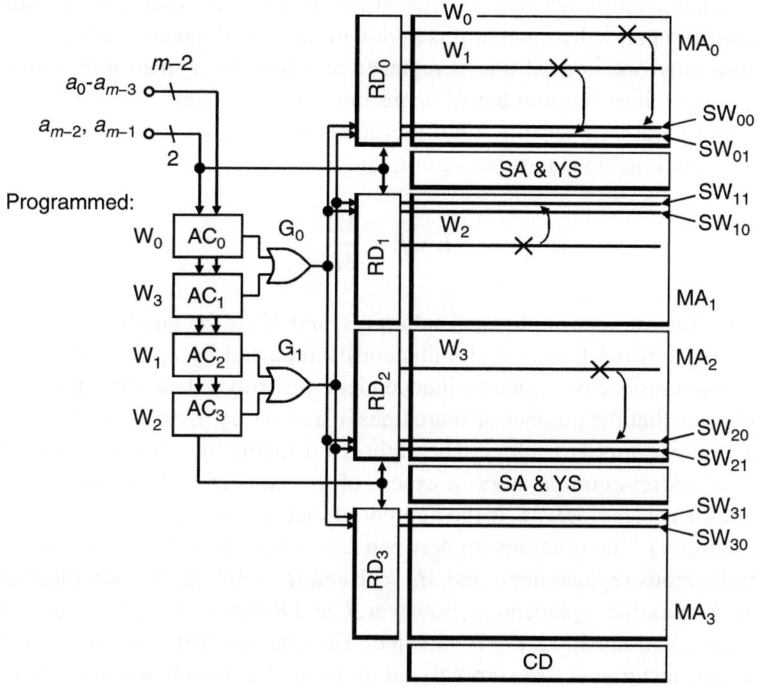

Fig. 2.30 Flexible intrasubarray replacement applied to a memory with array division [5, 20]

are not directly connected but through the OR gates G_0 and G_1. Since each address comparator compares both intra- and intersubarray address bits, the replacement unit is one wordline. Each logical-OR of the outputs of the two address comparators is commonly applied to all the subarrays. The intersubarray address bits select one of the four spare wordlines. In this example, the addresses of faulty normal wordlines W_0, W_1, W_2, and W_3 are programmed in address comparator AC_0, AC_2, AC_3, and AC_1, respectively. Wordlines W_0, W_1, W_2, and W_3 are thereby replaced by spare wordlines SW_{00}, SW_{01}, SW_{11}, and SW_{20}, respectively. This technique features a flexible relationship between spare lines and address comparators. In Figs. 2.28 and 2.29, this relationship is fixed so that a spare line can be activated only by a particular address comparator. For example, spare wordline SW_{20} in Fig. 2.28 is activated only by AC_0, and SW_{20} in Fig. 2.29 is activated only by AC_4. However, in Fig. 2.30, a spare line can be activated by one of several address comparators. For example, spare wordline SW_{20} can be activated by either AC_0 or AC_1 through OR gate G_0. In addition, an address comparator can activate one of several spare lines. For example, SC_0 can activate SW_{00}, SW_{10}, SW_{20}, or SW_{30}. This flexible relationship provides the following advantages. First, both spare-line usage efficiency and address comparator usage efficiency are good, while the former is poor and the latter is good in the simultaneous replacement (Fig. 2.28), and the former is good and the latter is poor in the individual replacement (Fig. 2.29). This enables smaller chip-area penalty due to redundancy. Second, the probability of unsuccessful repair is low, while it is high with the simultaneous replacement. This is because only one normal line is replaced at a time by a spare line. Third, more flexible selection of the number of spare lines in a subarray L and the number of address comparators R enable a more efficient redundancy circuit. Generally, the following relationship stands between L and R:

$$L \leqslant R \leqslant \frac{ML}{M_0}, \tag{2.17}$$

where M is the number of physical subarrays, and M_0 is the number of subarrays in which faulty normal lines are simultaneously replaced by spare lines. Therefore, M/M_0 is the number of logically independent subarrays. The left-hand inequality sign indicates that the number of spare lines in a subarray in excess of the number of address comparators is useless. The right-hand inequality sign indicates that the number of address comparators in excess of the number of logically independent spare lines is useless (ML/M_0 is the number of logically independent spare lines in a whole memory). The relationship between L and R is fixed: $M_0 = M$ and $R = L$ in the simultaneous replacement, and $M_0 = 1$ and $R = ML$ in the individual replacement. In the flexible replacement, however, L and R can be chosen independently as long as the relationship (2.17) is satisfied. The characteristics of the intrasubarray replacement techniques are summarized in Table 2.3, which also include those of intersubarray replacement described in the next section.

The flexible replacement can also be applied to column redundancy as shown in Fig. 2.31. The memory array is divided into four subarrays similar to Figs. 2.28–2.30. Each column selection line CSL transmits the output of the column decoder to column switches (see Fig. 2.24 in detail). There is a spare column composed of a spare column selection line (SCSL) and spare datalines SD_0–SD_3.

Table 2.3 Characteristics of intrasubarray and intersubarray replacement techniques

	Replacement unit	Replacement region	No. of address comparators R	Usage of spare lines	Usage of address comparators	Figure
Intrasubarray						
Simultaneous replacement	M lines	Subarray	L	Poor	Good	2.28
Individual replacement	One line	Subarray	ML	Fair	Poor	2.29
Flexible replacement	One line	Subarray	$L \leq R \leq ML$	Fair	Good	2.30, 2.31
Intersubarray						
Distributed spare lines	One line	Chip	L	Good	Good	2.39
Concentrated spare lines	One line	Chip	L'	Very good	Very good	2.40

M is the number of subarrays, L is the number of spare lines in a subarray, and L' is the number of spare line in the spare subarray

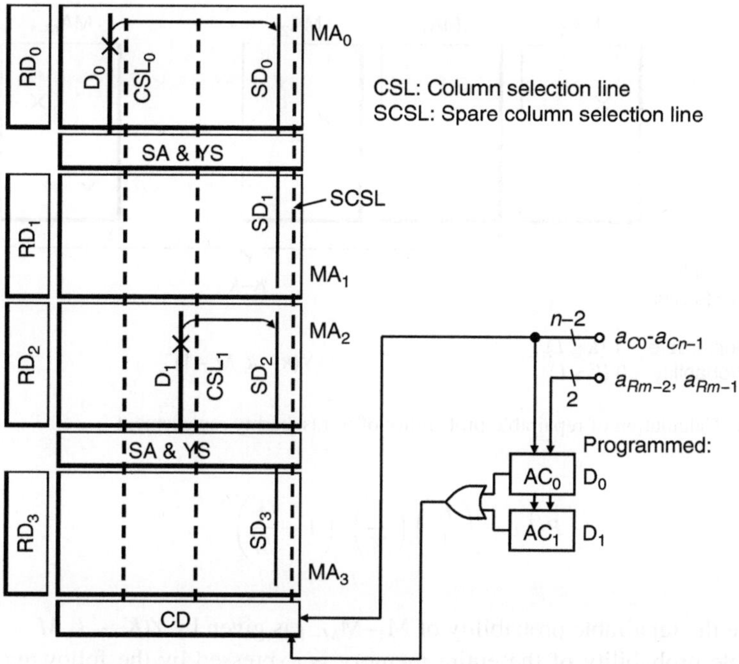

Fig. 2.31 Flexible intrasubarray replacement applied to column redundancy

The line SCSL is activated by the OR of two address comparators AC_0 and AC_1. Note that not only column address bits $a_{C_0} - a_{C_{n-1}}$ but also the upper two of row address bits, $a_{R_{m-2}}$ and $a_{R_{m-1}}$ (intersubarray address signals) are programmed in each comparator. In this example, the addresses of faulty normal datalines D_0 and D_1 are programmed in AC_0 and AC_1, respectively. Thus, the two datalines are respectively replaced by spare datalines SD_0 and SD_2 although there is only one spare column. Further improving the efficiency of column redundancy by using the results of row-address comparison is proposed in [28].

Let us calculate the repairable probability of the flexible intrasubarray replacement as a function of the number of faults. Note that the probability depends not only on the total number of faults but also on their distribution. Random fault distribution is assumed, and faults on spare lines and fatal faults are neglected here for simplicity. The repairable probability in the case of K faults being distributed in M subarrays is denoted as $Y(K, M)$. In the case of $M = 1$ (no array division), it is obvious that

$$Y(K,1) = 1(k \leqslant L \text{ and } k \leqslant R), \quad 0(k > L \text{ or } k > R). \tag{2.18}$$

Let us focus on a particular subarray M_0. If k faults are located in M_0, the remaining $(K - k)$ faults are in the other $(M - 1)$ subarrays M_1–M_{M-1} as shown in Fig. 2.32. The probability that k faults among K faults are located to M_0 is given by the binomial distribution:

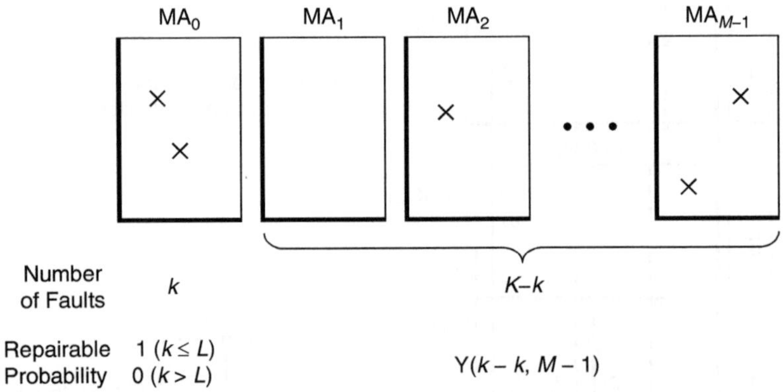

Fig. 2.32 Calculation of repairable probability of intrasubarray replacement

$$P(k) = \binom{K}{k}\left(\frac{1}{M}\right)^k\left(1-\frac{1}{M}\right)^{K-k}. \tag{2.19}$$

Since the repairable probability of M_1–M_{M-1} is given by $Y(K - k, M - 1)$, the repairable probability of the entire memory is expressed by the following recurrence formula:

$$\begin{aligned} Y(K,M) &= \sum_{k=0}^{\max(K,L)} P(k)Y(K-k,M-1) \\ &= \sum_{k=0}^{\max(K,L)} \binom{K}{k}\left(\frac{1}{M}\right)^k\left(1-\frac{1}{M}\right)^{K-k} \quad Y(K-k,M-1). \end{aligned} \tag{2.20}$$

In the case of $K > R$, however, $Y(K, M) = 0$. The calculated repairable probability is shown by the broken lines in Fig. 2.33. It gradually decreases with the increase in the number of faults K.

The calculated yield using a more practical model is shown in Fig. 2.34 [5], comparing the simultaneous replacement and flexible replacement. The yield improvement factors through both the replacement techniques are almost the same in a 4-Mbit DRAM. The advantage of the flexible replacement becomes apparent in 64-Mbit and 1-Gbit DRAMs, especially for a large defect density, that is, in the early stages of production.

A disadvantage of the flexible replacement is the problem of a "global" defect, that is, a defect causing two or more faults. Let us consider the multidivided dataline structure (Fig. 2.24). A defect on a dataline is "local" and produces no effect on the other subarrays. However, if a defect exists on a sense-amplifier, two datalines connected to the amplifier fail simultaneously. A defect on a column selection line causes all the data lines connected to the line to fail simultaneously. Thus these types of defects are global. In the case of the hierarchical wordline structure (Figs. 2.25 and 2.26),

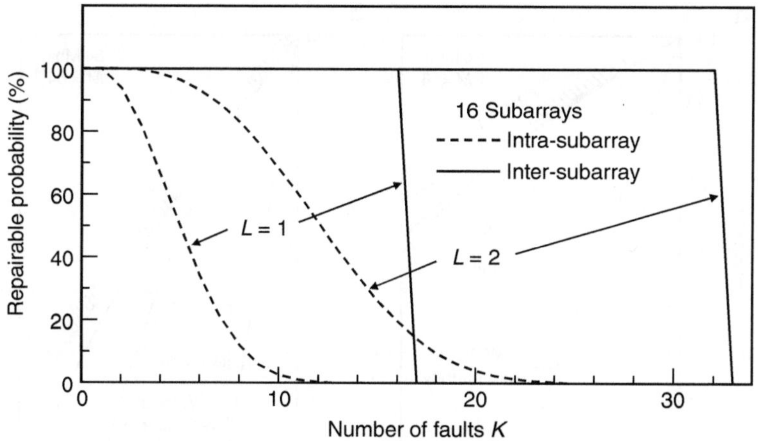

Fig. 2.33 Comparison between intra and intersubarray replacement. Reproduced from [20] with permission; © 2010 IEEE

a defect on a MWL is global. Since only one faulty normal line is replaced by a spare line in the flexible replacement, more than one address comparator is needed to repair the faults caused by a global defect. This may result in a deficiency of address comparators. Variable replacement unit using a ternary address comparator [5] can solve this problem. It features that not only ZERO and ONE, but also a "don't-care" value can be programmed. The don't-care value is assumed to coincide with both ZERO and ONE. Programming don't care in an address comparator specifies that the address bit is not compared with the input address. Using the don't care makes replacement unit (a group of memory cells replaced at a time) variable according to defect mode. For example, a sense-amplifier defect can be repaired by programming a don't care at the intersubarray address bit that specifies an upper/lower subarray of the sense amplifier ($a_{R_{m-2}}$ in Fig. 2.31), and by programming ZERO or ONE at the other address bits. This is because the intrasubarray address is common to the two faulty datalines. Table 2.4 shows the numbers of address comparators required to repair the various defects. A dataline or subwordline (local) defect requires an address comparator whether the replacement unit is fixed or variable. A global defect, however, requires two or more address comparators in the fixed replacement unit, while it requires only one in the variable replacement unit. Thus, the variable replacement unit reduces the number of address comparators required to repair the same number of defects and to achieve the same yield. A circuit diagram of the ternary address comparator is shown in Fig. 2.35. This circuit has two fuses, F_0 and F_1. Table 2.5 shows the programming method. ZERO or ONE is programmed by blowing one of the fuses as in Table 2.1. Don't care is programmed by remaining both fuses unblown, and the output c_i is always "1." Note the bottom two rows of the table. When both fuses are blown, the output c_i is always "0." This can be used to invalidate the address comparator when the corresponding spare line is found to be faulty. In this sense this circuit is "quaternary" rather than ternary.

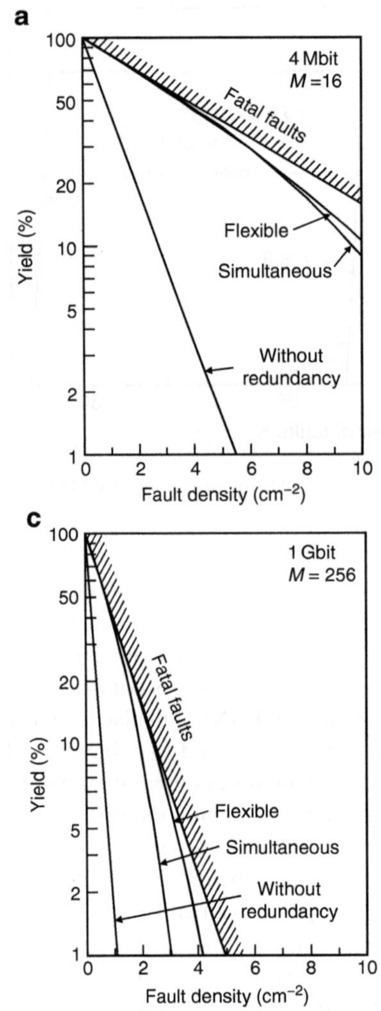

Fig. 2.34 Calculated yield using intrasubarray replacement redundancy techniques, simultaneous replacement ($L = R = 8$) and flexible replacement ($L = 4$, $R = 16$): (**a**) 4-Mbit DRAM (number of subarrays $M = 16$), (**b**) 64-Mbit DRAM ($M = 64$), and (**c**) 1-Gbit DRAM ($M = 256$). Reproduced from [5] with permission; © 2010 IEEE

Table 2.4 Number of address comparators required for repairing defects [5, 20]

Defect mode	No. of address comparators	
	Binary (fixed replacement unit)	Ternary (variable replacement unit)
Dataline	1	1
Sense amplifier	2	1
CSL	*M*	1
Sub wordline	1	1
Main wordline	*N*	1

M number of subarrays connected to a CSL, *N* number of sub wordlines connected to a main wordline

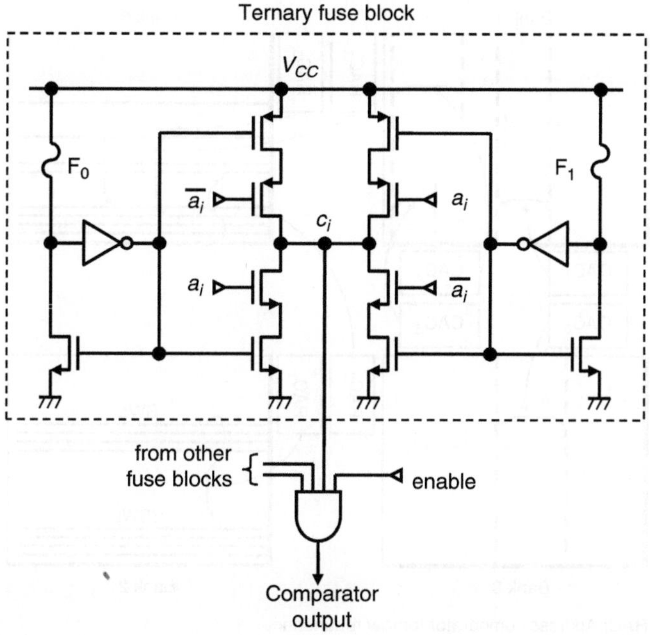

Fig. 2.35 Ternary address comparator for variable replacement unit. Reproduced from [5] with permission; © 2010 IEEE

Table 2.5 Truth table of ternary fuse block

Stored address	F_0	F_1	a_i	$\bar{a}_i$	c_i
Zero	Blown	Unblown	0	1	1
			1	0	0
One	Unblown	Blown	0	1	0
			1	0	1
Don't care	Unblown	Unblown	0	1	1
			1	0	1
Invalid	Blown	Blown	0	1	0
			1	0	0

2.5.3 Variations of Intrasubarray Replacement

It is possible to apply the intersubarray replacement described in the next section to memories with array divisions for further enhancing the replacement efficiency. In some cases, however, the intersubarray replacement can never be applied.

One of the cases is bank division. Some memories, such as synchronous DRAMs (SDRAMs) have a plurality of memory banks. Since different banks may be simultaneously active, interbank replacement (replacing a faulty normal line in a bank by a spare line in another bank) is not allowed. Figure 2.36 shows a double-data-rate (DDR) SDRAM with flexible intrabank replacement redundancy and variable replacement unit [29]. Address comparators for row redundancy RAC_0–RAC_3 are shared by four banks. Each comparator can be used for any

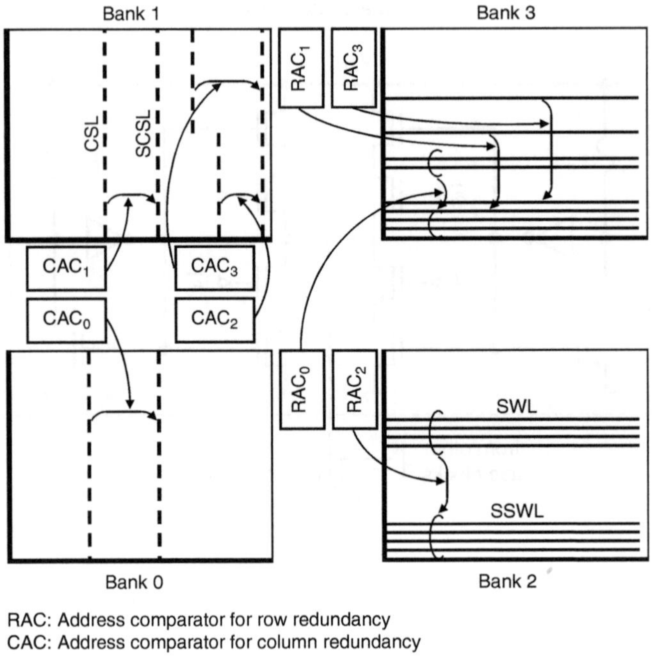

Fig. 2.36 Flexible intrasubarray replacement applied to a memory with bank division. Reproduced from [29] with permission; © 2010 IEEE

bank. In this case, RAC_0, RAC_1, and RAC_3 are used for replacing faulty wordlines in bank 3 by spare wordlines, and RAC_2 is used for bank 2. Similarly, address comparators for column redundancy CAC_0–CAC_3 are shared by two banks. Here, CAC_0 is used for bank 0 and CAC_1–CAC_3 are used for bank 1. In addition, the replacement unit is variable for both row and column redundancy. One, two, four, or eight SWLs can be replaced by spare wordlines at a time. A column-selection line (CSL) or a half of CSL can be replaced by a spare CSL.

Another case in which the intersubarray replacement is not allowed is multibit prefetch. Prefetching two or more bits simultaneously and outputting them serially are often used for increasing data-transfer rate. Figure 2.37 shows a DDR-SDRAM using 2-bit prefetch. Two subarrays store data of even and odd column addresses, respectively. The subarrays are simultaneously activated and two sets of data from them are serially outputted at the rising and falling edges of a clock signal. If the input column address is even (i.e., the lowest address bit $a_{C_0} = 0$), the activated column addresses in both subarrays are the same. However, if the input address is odd ($a_{C_0} = 1$), they are different; two bits of column address a_{C_1} and a_{C_2} in MA(even) equal to that in MA (odd) plus one. In this case, the two bits of input address are incremented by one and are supplied to MA(even). A straightforward approach is providing two sets of address comparators (including fuses), one for even and the other for odd. However, this doubles the circuit area. An approach to reduce the area penalty, even-odd sharing

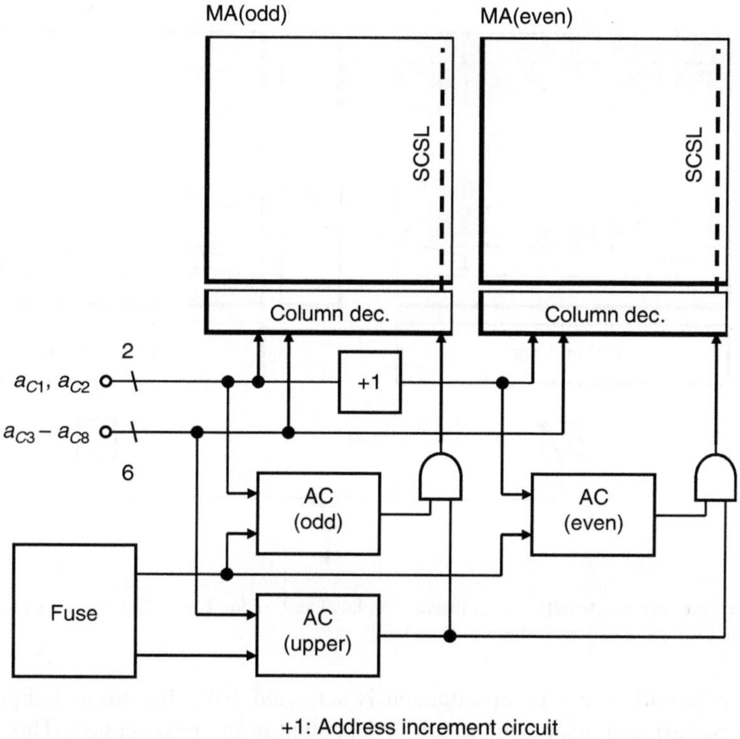

Fig. 2.37 Intrasubarray replacement applied to a memory using 2-bit prefetch [30]

scheme, is shown in Fig. 2.37 [30]. It features partial circuit sharing between the two subarrays. The lower-address (a_{C_1} and a_{C_2}) comparators for MA(even) and MA(odd) are separate, and the upper-address comparators and fuse sets are common. A spare column is activated by logical AND of the outputs of lower- and upper-address comparators. This scheme therefore halves the number of fuses required. In other words, if this approach has the same number of fuses as the straightforward approach, the number of spare columns can be doubled in order to improve the chip yield.

The access-time penalty due to redundancy is the delay time required for the address comparison as described in the previous section. Figure 2.38 shows a technique to eliminate this delay time for a high-speed SRAM [31]. In this technique, a faulty wordline in a subarray is replaced by a spare wordline in the adjacent subarray. The two subarrays are simultaneously activated and one of the data from them is selected according to the result of address comparison. This technique is difficult to be applied to row redundancy of general-purpose DRAMs because the dataline charging/discharging current is doubled. However, it is effective for DRAM column redundancy [32]. It is also reported that this technique was applied to the row redundancy of an ultrahigh-speed DRAM [33], in which power dissipation is not a major concern. Another disadvantage of this technique is that the number of spare lines is limited to one. In order to provide two spare lines, three

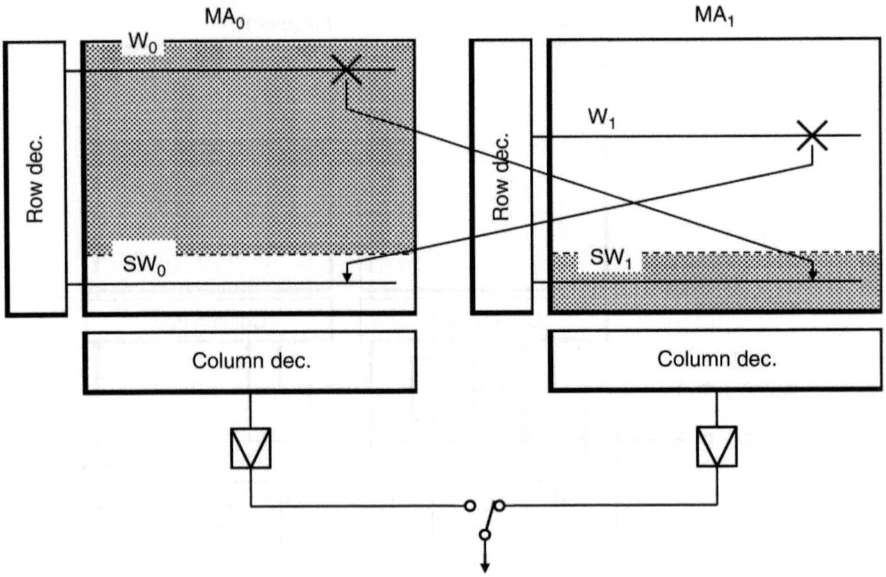

Fig. 2.38 No access-penalty intrasubarray replacement redundancy technique (simultaneous activation of normal and spare lines) [31, 33]

subarrays would have to be simultaneously activated. Note that this technique is not the intersubarray replacement which is described in the next section. This will be clear if the hatched areas in Fig. 2.38 are assumed to be a subarray and the nonhatched areas are assumed to be another subarray.

2.6 Intersubarray Replacement

With the further increase in memory-array division, the probability of clustered faults in a particular subarray becomes no more negligible. In the intrasubarray replacement, the number of spare lines in a subarray, L, must be larger or equal to the maximum number of faulty lines in each subarray to repair clustered faults. This causes the increase in L and chip-area penalty. To solve this problem, intersubarray replacement redundancy techniques [34–36] have been proposed, which permit a faulty normal line to be replaced by a spare line in any subarray. The replacement region is therefore an entire memory chip. They are classified into two categories.

In the distributed-spare-line approach [34] shown in Fig. 2.39, each subarray has its own spare lines like the intrasubarray replacement. Each spare line, however, can replace any faulty normal line not only in the same subarray but also in another subarray. Therefore at most LM faults clustered in a subarray can be repaired, where M is the number of subarrays. In this example, four clustered faulty normal wordlines W_0–W_3 are replaced by the spare wordlines in subarrays MA_0, MA_1, and MA_2. It is

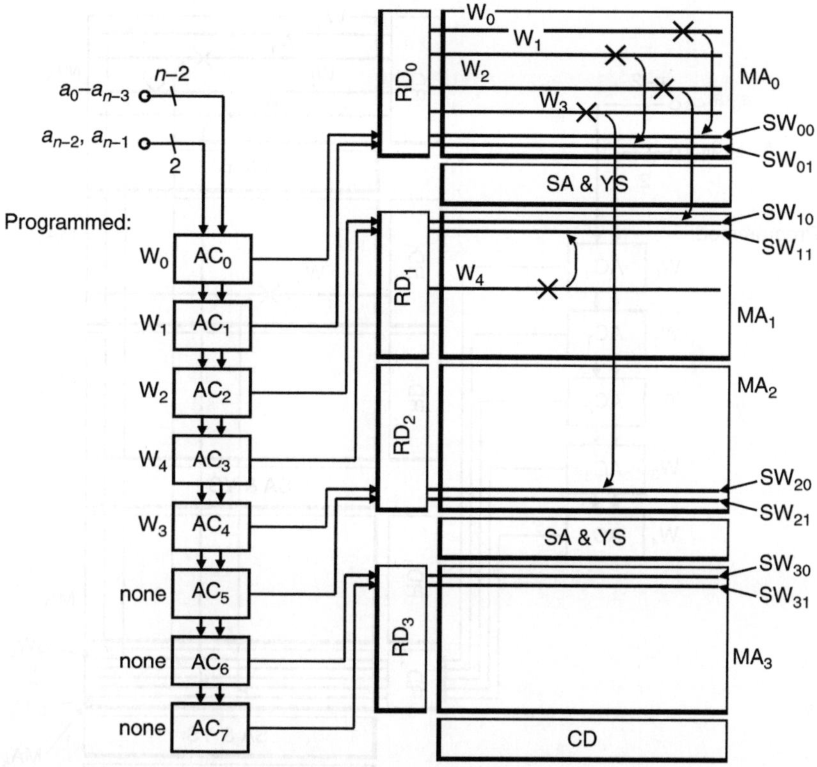

Fig. 2.39 Intersubarray replacement with distributed spare lines [34]

sufficient for successful repair that the number L is the average number of faulty lines in a subarray and is smaller than that of intrasubarray replacement. The number of address comparators R is equal to LM in this case. The number, however, can be reduced through the similar technique as the flexible replacement shown in Fig. 2.30.

In the concentrated-spare-line approach [35, 36] shown in Fig. 2.40, each subarray has no spare lines. There is a spare subarray MA_S, instead, composed of L' (here, $L' = 5$) spare lines. Each spare line can replace a faulty normal line in any subarray. Therefore, at most L' faults clustered in a subarray can be repaired. The number of address comparators R is equal to L'. This approach has an advantage of more flexible selection of $L'(= R)$ and better usage of address comparators compared to the distributed-spare-line approach. This is because the size of the spare subarray need not be the same as that of a normal subarray. The problem of this approach is that additional circuits (a decoder, a sense amplifier, etc.) for MA_S are needed. A solution of this problem using the hierarchical bitline architecture is proposed in [35].

Figure 2.33 compares the repairable probability using intra- and intersubarray replacement redundancy techniques [34, 36]. The probability for the former is calculated using (2.20). In the intrasubarray replacement, the repairable probability of a memory composed of M subarrays decreases with the increase in the number of

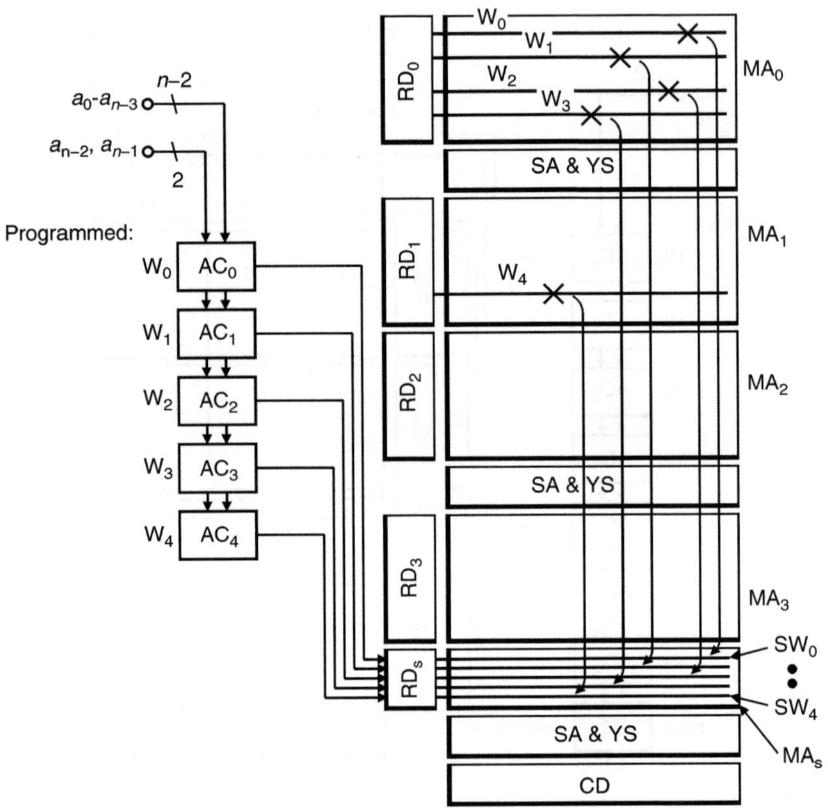

Fig. 2.40 Intersubarray replacement with concentrated spare lines [35, 36]

faults, K, because the probability of excessive ($>L$) faults in a particular subarray increases. On the other hand, the repairable probability is constantly 100% as long as $K \leq LM$ in the intersubarray replacement. When the number of subarrays is $M = 16$, the expectation of repairable faults with intersubarray replacement is about three times that with intrasubarray replacement.

The access-time penalty of the intersubarray replacement is usually larger than that of intrasubarray replacement. This is because not only an activated-line but also an activated subarray may be changed according to the result of address comparison.

2.7 Subarray Replacement

One of the problems with the increase in memory capacity is defects causing DC-characteristics faults, which mean the violation of DC specification of memories. In particular, excessive-standby-current (I_{SB}) is the most popular DC-characteristics

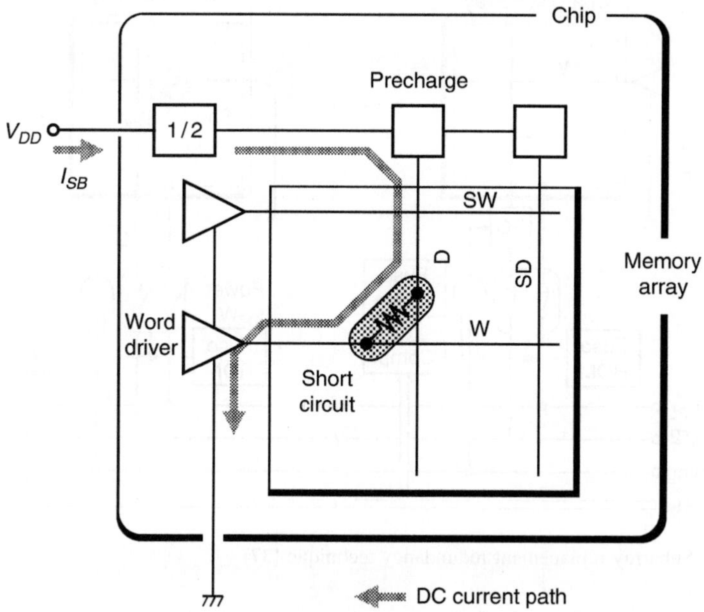

Fig. 2.41 Model of standby-current fault of a DRAM [37]

fault. An example of a defect that causes an I_{SB} fault of a DRAM is shown in Fig. 2.41 [37]. A short circuit between a wordline (electrically connected to the ground during standby state) and a dataline (connected to the dataline precharge voltage, $V_{DD}/2$) creates an illegal DC current path from $V_{DD}/2$ to ground. Replacing the wordline and dataline by a spare wordline and a spare dataline, respectively, inhibits the faulty lines from being accessed, but the current path still remains. Thus, the fault is not repaired by line replacement.

Several techniques have been reported to repair such faults. They include limiting the illegal current through the short circuit to a small value by a current limiter [36], cutting off the current path by a power switch controlled by a fuse [38]. Figure 2.42 shows another technique [37] where the replacement unit is a subarray. A subarray, including an I_{SB} fault, is replaced by an on-chip spare subarray. Each subarray has power switches for bitline precharge voltage $V_{DD}/2$ and memory-cell plate voltage V_{PL}, logic gates for timing signals, and a fuse to control them. The power switches of the faulty subarray are turned off and those of the spare subarray are turned on. Thus the I_{SB} fault is repaired by cutting of the DC current. The logic gates of the faulty subarray are also turned off to eliminate unnecessary power dissipation in the subarray. It is reported that this technique combined with the line replacement doubles the yield of a 256-Mbit DRAM [36]. One advantage of this technique is that the wordlines in an unused spare subarray can be used as spare wordlines of the concentrated-spare-line intersubarray replacement redundancy described in Sect. 2.6 [39].

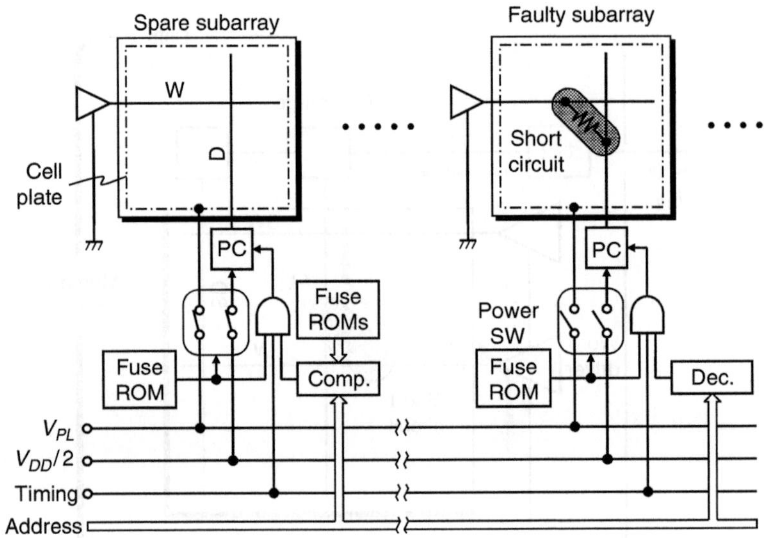

Fig. 2.42 Subarray-replacement redundancy technique [37]

Since this redundancy technique requires spare subarrays, it is not suitable for a small-capacity memory with a small number of subarrays. However, the number of subarrays rapidly increases with every generation as shown in Fig. 2.23. Therefore, the area penalty is allowable for DRAMs of 256 Mbit and beyond. It is interesting that the memory-array division, which was the barrier to the line-replacement redundancy, in turn, supports the subarray-replacement redundancy.

2.8 Devices for Storing Addresses

Redundancy techniques require devices for storing faulty addresses. These devices must be programmable as well as nonvolatile to retain the replacement information during power-off. The devices include fuses, antifuses, and nonvolatile memory cells. A fuse is initially in a conductive (low-resistance) state and reaches a nonconductive (high-resistance) state by programming. On the contrary, an antifuse is initially in a high-resistance state and reaches a low-resistance state. On the other hand, a nonvolatile memory cell exploits the change of its threshold voltage.

2.8.1 Fuses

The materials used for fuses include polysilicon, silicides ($MoSi_2$, $TaSi_2$), and metals (Cu, etc.). A fuse is blown by either electric current or laser and each has advantages and disadvantages.

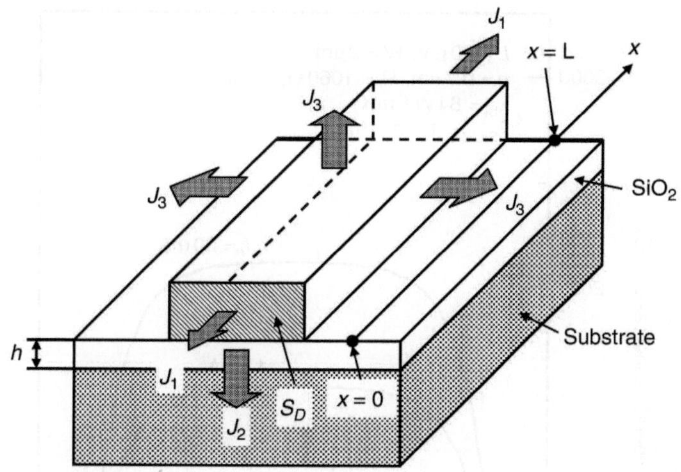

Fig. 2.43 Model of an electrical fuse and a coordinate used for one-dimensional thermal equation. Reproduced from [42] with permission; © 2010 JSAP

Electrical blowing requires no special equipment. In addition, faulty addresses can be programmed even after packaging (postpackaging programming) [40, 41]. However, it requires an on-chip programming circuit shown in Fig. 2.18, including a large transistor M_0 for driving a large blowing current. A special high-voltage V_{PP} is supplied from an extra pad probed at wafer test [15]. This voltage enhances the conductance of M_0 to supply a sufficient current. Whether the fuse is blown or not is determined by the address signal a_i according to the left half of Table 2.1. After programming, the V_{PP} pad is grounded by an on-chip pull-down circuit (not shown) to prevent inadvertent programming. A model of the electrical blowing of a polysilicon fuse is shown in Fig. 2.43 [42]. The Joule heat generated by the current and the fuse resistance diffuses along the fuse to the metal-polysilicon contacts at $x = 0$ and $x = L$ (J_1), to the substrate through SiO_2 (J_2), and to the surrounding glass layer (J_3). The temperature increase at the contacts and the substrate is assumed to be zero because the thermal conductivities of metal and Si are far larger than those of SiO_2 and glass. The fuse is heated up to the melting point (1,420°C) and is finally blown off. The temperature increase $T(x)$ at position x after a sufficiently long time (i.e., thermal equilibrium) is expressed as:

$$T(x) = T_0\left(1 - \frac{\sinh\frac{x}{L_0} + \sinh\frac{L-x}{L_0}}{\sinh\frac{L}{L_0}}\right), \tag{2.21}$$

where L is the fuse length, L_0 is the characteristic length determined by the thermal conductances of the polysilicon, SiO_2, and glass layer, and T_0 is the equilibrium temperature without J_1 [42]. The calculated temperature-increase distribution along the fuse is shown in Fig. 2.44. The fuse length should be designed to be at least several times of L_0 (here, $L_0 = 3.55$ μm).

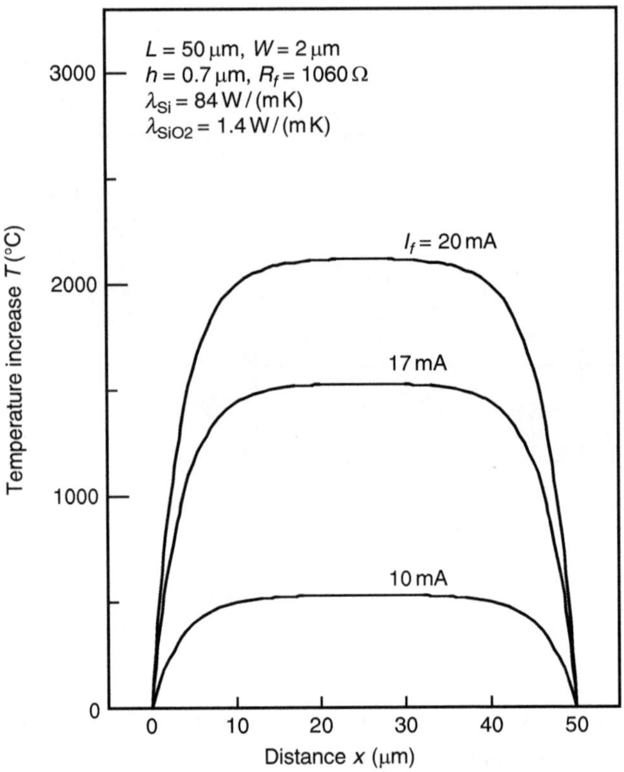

Fig. 2.44 Temperature distribution along an electrical fuse. Reproduced from [42] with permission; © 2010 JSAP

Blowing a fuse by laser requires a laser equipment. However, the area penalty is smaller because the on-chip programming circuit is not required. The area for fuses is determined by the diameter of laser spot and its alignment accuracy. It is therefore suitable for high-density memories with a large number of spare lines. However, the postpackaging programming is impossible.

Since the fuse and the surrounding layers are broken by blowing, contamination from the wreck is a major concern for both electrical blowing and laser blowing. Device structures for preventing the contamination are reported in [42, 43].

2.8.2 Antifuses

Various kind of antifuses have been proposed, including electrically destroying a p–n junction of a polysilicon diode [44], diffusing impurities into an intrinsic polysilicon by laser pulses [45], and electrically breaking down a dielectric [46]. Figure 2.45 shows the cross section of the antifuse for DRAM redundancy. It utilizes an oxide–nitride–oxide (ONO) film as a dielectric. This structure is compatible with the

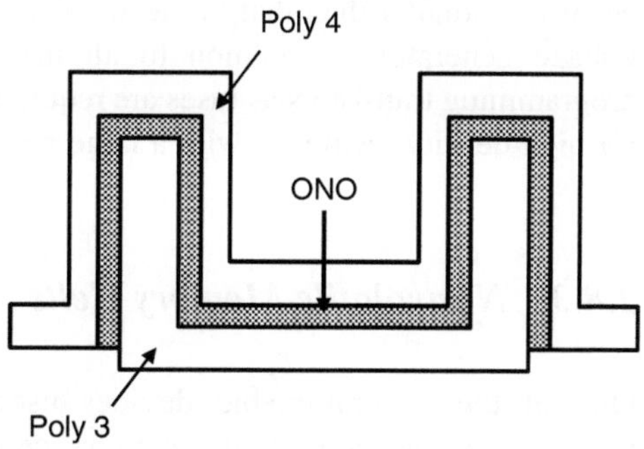

Fig. 2.45 Cross section of an antifuse using oxide–nitride–oxide (ONO) film for DRAM capacitor [46]

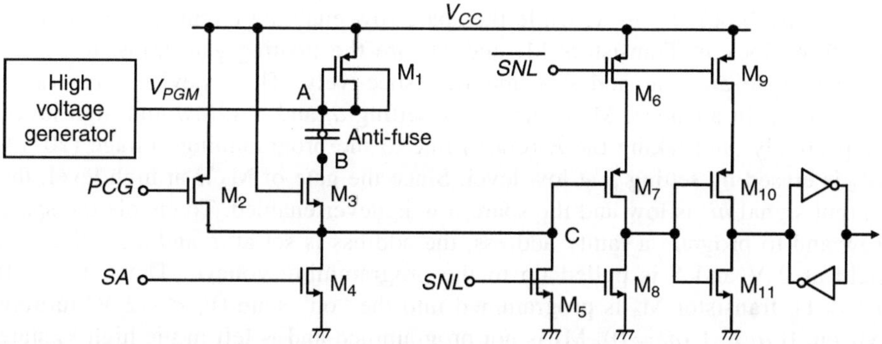

Fig. 2.46 Antifuse programming and sensing circuit. Reproduced from [46] with permission; © 2010 IEEE

memory cell of the DRAM and no additional process is required. Initially, the ONO film has an extremely high resistance (>100 MΩ). By applying a high-programming voltage (≥8 V), the film destructively breaks down and produces a short circuit between the polysilicon layers with a significantly lower resistance (<10 kΩ). The programming and sensing circuit of the antifuse is shown in Fig. 2.46. The programming procedure is as follows. First, signal PCG is low and node B is precharged to $V_{CC} - V_t$ (V_t is the threshold voltage of M_3) through M_2 and M_3 to prevent unselected antifuses from being programmed. The programming voltage V_{PGM} is generated by an on-chip high-voltage generator and is applied to all antifuses. Next, signal SA of selected antifuses is set high to discharge node B through M_3 and M_4 and to apply the full V_{PGM}, while the voltage across the unselected antifuses is limited. Note that M_3 protects M_5 and M_7 from the high voltage. In order to sense the state of the antifuse, signals SNL and $\overline{SNL}$ are set high and low, respectively. The voltage of node C is $V_{CC} - V_t$ if the antifuse is programmed, or zero if not. This analog voltage is sensed and latched by the subsequent circuit.

This technique features that the postpackaging programming is possible [46, 47] like the electrical fuses. Although the programming circuit is required, its area penalty is smaller than that of electrically blowing fuses. This is because the high-voltage generator is common to all the antifuses, while as many numbers of programming transistors as fuses are required. Therefore, this technique is promising for high-density memories with a large number of spares.

2.8.3 *Nonvolatile Memory Cells*

One of the programmable devices used for the redundancy of nonvolatile memories is the nonvolatile memory cell itself. Figure 2.47 shows an address comparator (corresponding to the address comparator in Fig. 2.17) used for row redundancy of an EEPROM [48]. The address programming circuit corresponds to the fuse block 0 – n–1, while the spare-row enabling circuit corresponds to the fuse block n. Transistors M_1 and M_4 are the floating gate transistors with tunnel oxide areas at nodes A and B, respectively. These devices are erased ($V_t > 6$ V) in advance. M_4 is erased by setting $\bar{a}_i$ and S at low and high level, respectively, and taking the E (erase) line to the programming voltage (20 V). M_1 is erased by setting S at low level. Since the gate of M_2 is at high level, the output signal *hit* is low and the spare row is never enabled. To enable the spare row and to program a faulty address, the address is set at a_i and $\bar{a}_i$, while E is held at 0 V and S is pulled up to the programming voltage. Thus, if $a_i = 0$ ($\bar{a}_i = 1$), transistor M_4 is programmed into the "on" state ($V_t < -2$ V) turning M_6 on. If $a_i = 1$ ($\bar{a}_i = 0$), M_4 is not programmed and is left in the high V_t state turning M_5 on. Concurrently, M_1 is also programmed into a low V_t state and M_2 is turned off. If the input address coincides to the programmed address, all the

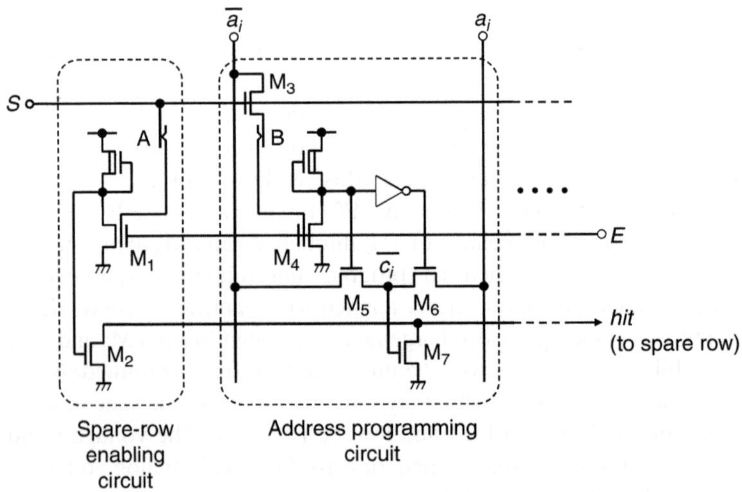

Fig. 2.47 Address comparator using EEPROM cells as programmable elements. Reproduced from [48] with permission; © 2010 IEEE

Table 2.6 Truth table of the address comparator using EEPROM cells

Erasing (E = high)			Programming (E = low)			Normal operation		
a_i	$\bar{a}_i$	M4	a_i	$\bar{a}_i$	M4	a_i	$\bar{a}_i$	$\bar{c}_i$
1	0	Erased (high V_t)	0	1	Programmed (low V_t)	0	1	0
						1	0	1
			1	0	Not programmed (high V_t)	0	1	1
						1	0	0

pull-down transistors (M_2, M_7) are turned off and the output signal *hit* is pulled up to a high level by a pull-up circuit (not shown), thus enabling the spare row. The operations of the circuit are summarized in Table 2.6. This technique features that postpackaging programming is possible like the antifuses. A nonvolatile memory cell fabricated with standard CMOS process and its application to memory redundancy are also reported [49, 50].

2.9 Testing for Redundancy

Testing for memories with redundancy includes the following three categories:

1. Normal-element test: testing normal elements and determining which normal elements should be replaced by which spare elements. This test is essential for redundancy techniques.
2. Spare-element test: checking whether each spare element that replaces a faulty normal element is faulty or not in advance. This test is optional but is effective for reducing the probability of unsuccessful repair caused by faulty spare elements.
3. Postrepair diagnostics: checking after packaging whether a memory LSI is a perfect chip or a repaired chip (signature), and if the latter, which addresses are repaired (roll-call). This test enables chip diagnostics by the manufacturer.

The normal-element test is usually performed by a memory tester. A memory tester has a fail-bit memory (FBM) for recording the pass/fail of each bit of the memory under test. After scanning an entire memory, a replacing solution is determined by analyzing the failure pattern in the FBM. Faulty rows must be replaced by spare rows and faulty columns must be replaced by spare columns. On the other hand, faulty bits can be replaced by either spare rows or spare columns. Therefore, a decision algorithm for finding a replacement solution is needed. The algorithm should meet the following requirements. First, the probability of overlooking (being unable to find a replacement solution though it does exist) should be minimized to maximize yield. Second, the decision time should be minimized in order not to occupy an expensive memory tester for a long time. In particular, a nonrepairable chip should be aborted as early as possible. Some repair algorithms to make the decisions are proposed to find an optimal replacement solution [51–53]. Figure 2.48 shows an algorithm using row/column fault counters. After testing an

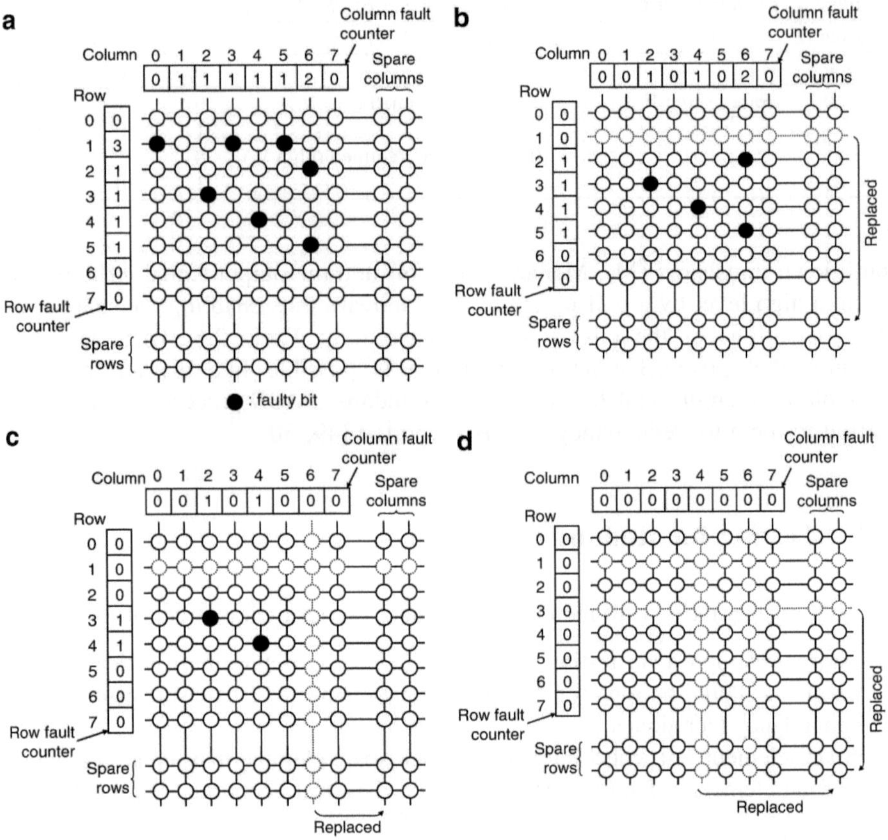

Fig. 2.48 Replacement algorithm: (**a**) just after testing an entire memory, (**b**) row one is replaced by a spare row, (**c**) column six is replaced by a spare column, and (**d**) row three and column four are replaced by spares

entire memory, the number of faulty bits on each row/column is recorded in each row/column fault counter (Fig. 2.48a). The algorithm consists of three steps.

Step 1: If the total number of faulty bits exceeds $N_X R_Y + N_Y R_X - R_X R_Y$, the chip cannot be repaired and is aborted, where N_X, N_Y, R_X, and R_Y are the number of normal rows, normal columns, spare rows, and spare columns, respectively. Since the example of Fig. 2.48a does not meet this condition, we go to the next step.

Step 2: A row having more faulty bits than the number of remaining spare columns must be replaced by a spare row, and a column having more faulty bits than the number of remaining spare rows must be replaced by a spare column. This step is repeated until no rows/columns meet the condition. In this example, row 1 has three faults and replaced by a spare row. The number of remaining spare rows becomes 1 and the contents of the counters are revised (Fig. 2.48b).

Step 3: A row or a column with maximum number of faulty bits is replaced by a spare row or a spare column. This step is repeated until all faulty bits are repaired (success) or spares are exhausted (failure). In this example, column 6 is first replaced by a spare column because it has two faults (Fig. 2.48c). Next row 3 and column 4 (or row 4 and column 2) are replaced by spares (Fig. 2.48d).

This algorithm is sufficiently good for usual faulty-bit distribution patterns though it cannot find a solution for some patterns [52]. Build-in self-test (BIST) and build-in self-repair (BISR) techniques utilizing on-chip circuitry instead of testers as well as repair algorithms suitable for them are also reported for test-cost reduction [54, 55].

Figure 2.49 shows a circuit for the spare-element test [56]. Before blowing fuses, each signal $\overline{hit_i}$ (the output of each address comparator) is high. The test circuit is enabled by setting the test pad at high level. During the test mode, any spare row can be selected by the row address signals a_0–a_2 and the memory cells connected to the selected spare row can be written and read. Thus, this circuit allows testing the spare rows without fuse programming. After the test, fuses are blown if necessary. During normal operation (test pad is low), each spare row is controlled by the signal $\overline{hit_i}$.

Figure 2.50a shows a DRAM with the postrepair diagnostics [57], which is realized by adding the extra circuitry within the dashed line. An additional fuse is blown only in the case of a repaired chip to enable the comparator and the roll-call decoder. The operating waveforms of row-redundancy diagnostics are shown in Fig. 2.50b. They are the same as those of "$\overline{\text{RAS}}$-only refresh" except that a super high-level voltage (8–10 V) is applied to the data input terminal D_{IN}. A row address A_R is specified at the falling edge of $\overline{\text{RAS}}$. The result of the test is available at the data output terminal D_{OUT}. If the device under test is a repaired chip, D_{OUT} goes high or low according to A_R being one of the repaired addresses or not. On the other hand, D_{OUT} remains high impedance in the case of a perfect chip. Column redundancy diagnostics are performed with waveforms like "early write" operation.

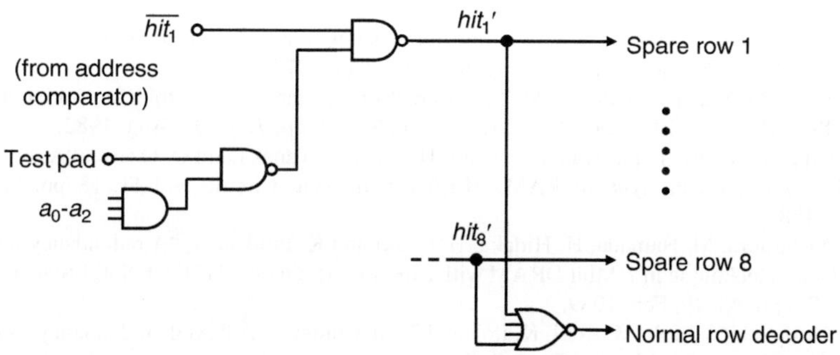

Fig. 2.49 Circuit for testing spare elements. Reproduced from [56] with permission; © 2010 IEEE

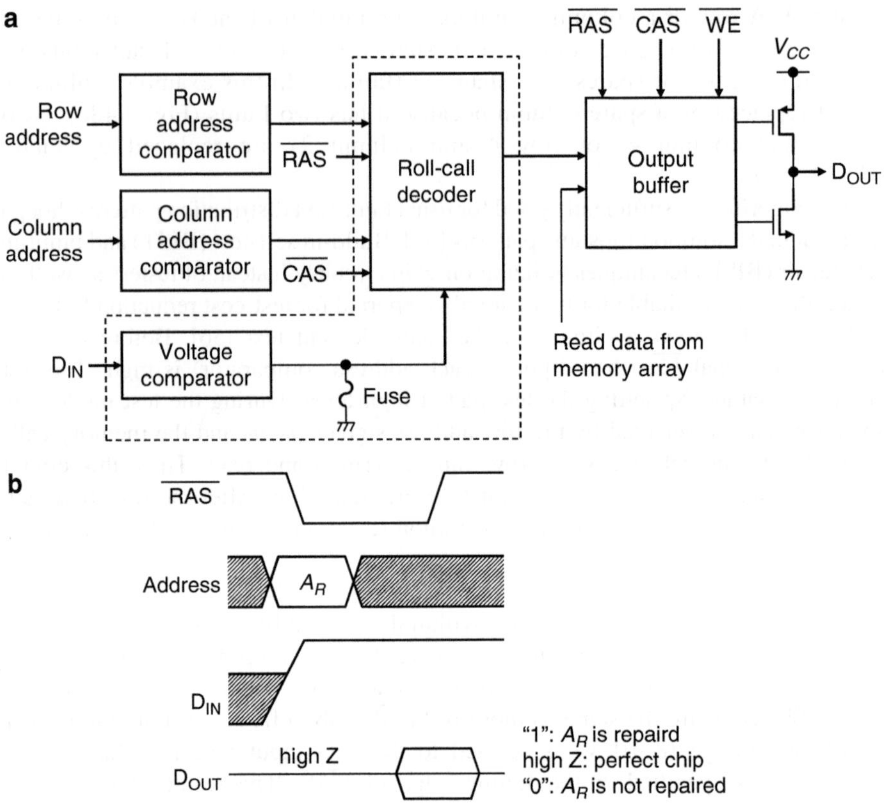

Fig. 2.50 Redundancy diagnostics circuit: (**a**) circuit diagram and (**b**) operating waveforms of row-redundancy diagnostics. Reproduced from [57] with permission; © 2010 IEEE

References

1. S. E. Schuster, "Multiple word/bit line redundancy for semiconductor memories," IEEE J. Solid-State Circuits, vol. SC-13, pp. 698–703, Oct. 1978.
2. T. Mano, M. Wada, N. Ieda and M. Tanimoto, "A redundancy circuit for a fault-tolerant 256K MOS RAM," IEEE J. Solid-State Circuits, vol. SC-17, pp. 726–731, Aug. 1982.
3. S. Fujii, K. Natori, T. Furuyama, S. Saito, H. Toda, T. Tanaka and O. Ozawa, "A low-power sub 100 ns 256K bit dynamic RAM," IEEE J. Solid-State Circuits, vol. SC-18, pp. 441–446, Oct. 1983.
4. Y. Nishimura, M. Hamada, H. Hidaka, H. Ozaki and K. Fujishima, "A redundancy test-time reduction technique in 1-Mbit DRAM with a multibit test mode," IEEE J. Solid-State Circuits, vol. 24, pp. 43–49, Feb. 1989.
5. M. Horiguchi, J. Etoh, M. Aoki, K. Itoh and T. Matsumoto, "A flexible redundancy technique for high-density DRAMs," IEEE J. Solid-State Circuits, vol. 26, pp. 12–17, Jan. 1991.
6. C. H. Stapper, Jr., "On a composite model to the IC yield problem," IEEE J. Solid-State Circuits, vol. SC-10, pp. 537–539, Dec. 1975.

7. C. H. Stapper, A. N. McLaren and M. Dreckmann, "Yield model for productivity optimization of VLSI memory chips with redundancy and partially good product," IBM J. Res. Dev., vol. 24, pp. 398–409, May 1980.
8. T. Okabe, M. Nagata and S. Shimada, "Analysis on yield of integrated circuits and a new representation for the yield," Trans. IEE J., vol, 92-C, pp. 399–406, Dec. 1972 (in Japanese).
9. C. H. Stapper, "Yield model for fault clusters within integrated circuits," IBM J. Res. Dev., vol. 28, pp. 636–640, Sep. 1984.
10. S. Kikuda, H. Miyamoto, S. Mori, M. Niiro and M. Yamada, "Optimized redundancy selection based on failure-related yield model for 64-Mb DRAM and beyond," IEEE J. Solid-State Circuits, vol. 26, pp. 1550–1555, Nov. 1991.
11. T. Yamagata, H. Sato, K. Fujita, Y. Nishimura and K. Anami, "A distributed globally replaceable redundancy scheme for sub-half-micron ULSI memories and beyond," IEEE J. Solid-State Circuits, vol. 31, pp. 195–201, Feb. 1996.
12. K. Imamiya, J. Miyamoto, N. Ohtsuka, N. Tomita and Y. Iyama, "Statistical memory yield analysis and redundancy design considering fabrication line improvement," IEICE Trans. Electron., vol. E76-C, pp. 1626–1631, Nov. 1993.
13. R. P. Cenker, D. G. Clemons, W. R. Huber, J. B. Petrizzi, F. J. Procyk and G. M. Trout, "A fault-tolerant 64K dynamic random-access memory," IEEE Trans. Electron Devices, vol. ED-26, pp. 853–860, June 1979.
14. E. A. Reese, D. W. Spaderna, S. T. Flannagan and F. Tsang, "A 4K × 8 dynamic RAM with self-refresh," IEEE J. Solid-State Circuits, vol. SC-16, pp. 479–487, Oct. 1981.
15. K. Kokkonen, P. O. Sharp, R. Albers, J. P. Dishaw, F. Louie and R. J. Smith, "Redundancy techniques for fast static RAMs," in ISSCC Dig. Tech. Papers, Feb. 1981, pp. 80–81.
16. A. Ohba, S. Ohbayashi, T. Shiomi, S. Takano, K. Anami, H. Honda, Y. Ishigaki, M. Hatanaka, S. Nagao and S. Kayano, "A 7-ns 1-Mb BiCMOS ECL SRAM with shift redundancy," IEEE J. Solid-State Circuits, vol. 26, pp. 507–512, Apr. 1991.
17. H. Noda, K. Inoue, M. Kuroiwa, A. Amo, A. Hachisuka, H. J. Mattausch, T. Koide, S. Soeda, K. Dosaka and K. Arimoto, "A 143MHz 1.1W 4.5Mb dynamic TCAM with hierarchical searching and shift redundancy architecture," ISSCC Dig. Tech. Papers, Feb. 2004, pp. 208–209.
18. A. Roth, D. Foss, R. McKenzie and D. Perry, "Advanced ternary CAM circuits on 0.13 μm logic process technology," in Proc. CICC, Oct. 2004, pp. 465–468.
19. T. Namekawa, S. Miyano, R. Fukuda, R. Haga, O. Wada, H. Banba, S. Takeda, K. Suda, K. Mimoto, S. Yamaguchi, T. Ohkubo, H. Takato and K. Numata, "Dynamically shift-switched dataline redundancy suitable for DRAM macro with wide data bus," IEEE J. Solid-State Circuits, vol. 35, pp. 705–712, May 2000.
20. M. Horiguchi, "Redundancy techniques for high-density DRAMs," in Proc. Int. Conf. on Innovative Systems Silicon, Oct. 1997, pp. 22–29.
21. R. Hori, K. Itoh, J. Etoh, S. Asai, N. Hashimoto, K. Yagi and H. Sunami, "An experimental 1 Mbit DRAM based on high S/N design," IEEE J. Solid-State Circuits, vol. SC-19, pp. 634–640, Oct. 1984.
22. K. Itoh, "Trends in megabit DRAM circuit design," IEEE J. Solid-State Circuits, vol. 25, pp.778–789, June 1990.
23. K. Itoh, *VLSI Memory Design*, Baifukan, Tokyo, 1994 (in Japanese), Chapter 2.
24. K. Itoh, *VLSI Memory Chip Design*, Springer, NY, 2001, Chapter 3.
25. M. Yoshimoto, K. Anami, H. Shinohara, T. Yoshihara, H. Takagi, S. Nagao, S. Kayano and T. Nakano, "A divided word-line structure in the static RAM and its application to a 64k full CMOS RAM," IEEE J. Solid-State Circuits, vol. SC-18, pp. 479–485, Oct. 1983.
26. K. Noda, T. Saeki, A. Tsujimoto, T. Murotani and K. Koyama, "A boosted dual word-line decoding scheme for 256Mb DRAMs," in Symp. VLSI Circuits Dig. Tech. Papers, June 1992, pp. 112–113.
27. D. Galbi, K. Althoff, R. Parent, O. Kiehl, R. Houghton, F. Bonner, M. Killian, A. Wilson, K. Lau, M. Clinton, D. Chapman and H. Fischer, "A 33-ns 64-Mbit DRAM with master-wordline architecture," in Proc. ESSCIRC, Sep. 1992, pp. 131–134.

28. K. Furutani, T. Hamamoto, T. Miki, M. Nakano, T. Kono, S. Kikuda, Y. Konishi and T. Yoshihara, "Highly flexible row and column redundancy and cycle time adaptive read data path for double data rate synchronous memories," IEICE Trans. Electron., vol. E88-C, pp. 255–263, Feb. 2005.
29. Y. Takai, M. Fujita, K. Nagata, S. Isa, S. Nakazawa, A. Hirobe, H. Ohkubo, M. Sakao, S. Horiba, T. Fukase, Y. Takaishi, M. Matsuo, M. Komuro, T. Uchida, T. Sakoh, K. Saino, S. Uchiyama, Y. Takada, J. Sekine, N. Nakanishi, T. Oikawa, M. Igeta, H. Tanabe, H. Miyamoto, T. Hashimoto, H. Yamaguchi, K. Koyama, Y. Kobayashi and T. Okuda, "A 250-Mb/s/pin, 1-Gb double-data-rate SDRAM with a bidirectional delay and an interbank shared redundancy scheme," IEEE J. Solid-State Circuits, vol. 35, pp. 149–162, Feb. 2000.
30. H. Yahata, Y. Okuda, H. Miyashita, H. Chigasaki, B. Taruishi, T. Akiba, Y. Kawase, T. Tachibana, S. Ueda, S. Aoyama, A. Tsukimori, K. Shibata, M. Horiguchi, Y. Saiki and Y. Nakagome, "A 256-Mb double-data-rate SDRAM with a 10-mW analog DLL circuit," in Symp. VLSI Circuits Dig. Tech. Papers, June 2000, pp. 74–75.
31. K. Sasaki, K. Ishibashi, T. Yamanaka, N. Hashimoto, T. Nishida, K. Shimohigashi, S. Hanamura and S. Honjo, "A 9-ns 1-Mbit CMOS SRAM," IEEE J. Solid-State Circuits, vol. 24, pp. 1219–1225, Oct. 1989.
32. H. Yamauchi, T. Suzuki, A. Sawada, T. Iwata, T. Tsuji, M. Agata, T. Taniguchi, Y. Odake, K. Sawada, T. Ohnishi, M. Fukumoto, T. Fujita and M. Inoue, "A circuit technology for high-speed battery-operated 16-Mb CMOS DRAM's," IEEE J. Solid-State Circuits, vol. 28, pp. 1084–1091, Nov. 1993.
33. Y. Yokoyama, N. Itoh, M. Katayama, M. Hasegawa, K. Takashima, H. Akasaki, M. Kaneda, T. Ueda, Y. Tanaka, E. Yamasaki, M. Todokoro, K. Toriyama, H. Miki, M. Yagyu, T. Kobayashi, S. Miyaoka and N. Tamba, "A 1.8-V embedded 18-Mb DRAM macro with a 9-ns RAS access time and memory-cell area efficiency of 33%," IEEE J. Solid-State Circuits, vol. 36, pp. 503–509, Mar. 2001.
34. K. Ishibashi, K. Komiyaji, S. Morita, T. Aoto, S. Ikeda, K. Asayama, A. Koike, T. Yamanaka, N. Hashimoto, H. Iida, F. Kojima, K. Motohashi and K. Sasaki, "A 12.5-ns 16-Mb CMOS SRAM with common-centroid-geometry-layout sense amplifiers," IEEE J. Solid-State Circuits, vol. 29, pp. 411–418, Apr. 1994.
35. M. Asakura, T. Oishi, S. Tomishima, H. Hidaka, K. Arimoto and K. Fujishima, "A hierarchical bit-line architecture with flexible redundancy and block compare test for 256Mb DRAM," in Symp. VLSI Circuits Dig. Tech. Papers, May 1993, pp. 93–94.
36. T. Kirihata, Y. Watanabe, H. Wong, J. K. DeBrosse, M. Yoshida, D. Katoh, S. Fujii, M. R. Wordeman, P. Poechmueller, S. A. Parke and Y. Asao, "Fault-tolerant designs for 256 Mb DRAM," IEEE J. Solid-State Circuits, vol. 31, pp. 558–566, Apr. 1996.
37. G. Kitsukawa, M. Horiguchi, Y. Kawajiri, T. Kawahara, T. Akiba, Y. Kawase, T. Tachibana, T. Sakai, M. Aoki, S. Shukuri, K. Sagara, R. Nagai, Y. Ohji, N. Hasegawa, N. Yokoyama, T. Kisu, H. Yamashita, T. Kure and T. Nishida, "256-Mb DRAM circuit technologies for file applications," IEEE J. Solid-State Circuits, vol. 28, pp. 1105–1113, Nov. 1993.
38. K. Furutani, T. Ooishi, M. Asakura, H. Hidaka, H. Ozaki and M. Yamada, "A board level parallel test circuit and a short circuit failure repair circuit for high-density, low-power DRAMs," IEICE Trans. Electron., vol. E80-C, pp. 582–589, Apr. 1997.
39. M. Asakura, T. Ohishi, M. Tsukude, S. Tomishima, H. Hidaka, K. Arimoto, K. Fujishima, T. Eimori, Y. Ohno, T. Nishimura, M. Yasunaga, T. Kondoh, S. Satoh, T. Yoshihara and K. Demizu, "A 34ns 256Mb DRAM with boosted sense-ground scheme," in ISSCC Dig. Tech. Papers, Feb. 1994, pp. 140–141.
40. K. Lim, S. Kang, J. Choi, J. Joo, Y. Lee, J. Lee S. Cho and B. Ryu, "Bit line coupling scheme and electrical fuse circuit for repairable operation of high density DRAM," Symp. VLSI Circuits Dig. Tech. Papers, June 2001, pp. 33–34.
41. K. H. Kyung, C. W. Kim, J. Y. Lee, J. H. Kook, S. M. Seo, D. Y. Kim, J. H. Kim, J. Sunwoo, H. C. Lee, C. S. Kim, B. H. Jeong, Y. S. Sohn, S. P. Hong, J. H. Lee, J. H. Yoo and S. I. Cho, "A 800Mb/s/pin 2Gb DDR2 SDRAM with an 80nm triple metal technology," ISSCC Dig. Tech. Papers, Feb. 2005, pp. 468–469.

42. K. Shimohigashi, M. Ishihara and S. Shimizu, "Redundancy techniques for dynamic RAMs," Jpn. J. Appl. Phys., vol. 22, pp. 63–67, Apr. 1983.
43. S. Ohbayashi, M. Yabuuchi, K. Kono, Y. Oda, S. Imaoka, K. Usui, T. Yonezu, T. Iwamoto, K. Nii, Y. Tsukamoto, M. Arakawa, T. Uchida, M. Okada, A. Ishii, T. Yoshihara, H. Makino, K. Ishibashi and H. Shinohara, "A 65nm embedded SRAM with wafer level burn-in mode, leak-bit redundancy and Cu e-trim fuse for known good die," IEEE J. Solid-State Circuits, vol. 43, pp. 96–108, Jan. 2008.
44. T. Mano, K. Takeya, T. Watanabe, N. Ieda, K. Kiuchi, E. Arai, T. Ogawa and K. Hirata, "A fault-tolerant 256K RAM fabricated with molybdenum-polysilicon technology," IEEE J. Solid-State Circuits, vol. SC-15, pp. 865–872, Oct. 1980.
45. O. Minato, T. Masuhara, T. Sasaki, Y. Sakai, T. Hayashida, K. Nagasawa, K. Nishimura and T. Yasui, "A Hi-CMOSII 8K $\times$ 8 bit static RAM," IEEE J. Solid-State Circuits, vol. SC-17, pp. 793–798, Oct. 1982.
46. J.-K. Wee, W. Yang, E.-K. Ryou, J.-S. Choi, S.-H. Ahn, J.-Y. Chung and S.-C. Kim, "An antifuse EPROM circuitry scheme for field-programmable repair in DRAM," IEEE J. Solid-State Circuits, vol. 35, pp. 1408–1414, Oct. 2000.
47. J.-K. Wee, K.-S. Min, J.-T. Park, S.-P. Lee, Y.-H. Kim, T.-H. Yang, J.-D. Joo and J.-Y. Chung, "A post-package bit-repair scheme using static latches with bipolar-voltage programming antifuse circuit for high-density DRAMs," IEEE J. Solid-State Circuits, vol. 37, pp. 251–254, Feb. 2002.
48. E. M. Lucero, N. Challa and J. Fields Jr., "A 16 kbit smart 5 V-only EEPROM with redundancy," IEEE J. Solid-State Circuits, vol. SC-18, pp. 539–544, Oct. 1983.
49. S. Shukuri, K. Yanagisawa and K. Ishibashi, "CMOS process compatible ie-flash (inverse gate electrode flash) technology for system-on-a chip," IEICE Trans. Electron., vol. E84-C, pp. 734–739, June 2001.
50. M. Yamaoka, K. Yanagisawa, S. Shukuri, K. Norisue and K. Ishibashi, "A system LSI memory redundancy technique using an ie-flash (inverse-gate-electrode flash) programming circuit," IEEE J. Solid-State Circuits, vol. 37, pp. 599–604, May 2002.
51. M. Tarr, D. Boudreau and R. Murphy, "Defect analysis system speeds test and repair of redundant memories," Electronics, vol. 57, pp. 175–179, Jan. 1984.
52. J. R. Day, "A fault-driven, comprehensive redundancy algorithm," IEEE Design Test Comput., vol. 2, pp. 35–44, June 1985.
53. W. K. Huang, Y.-N. Shen and F. Lombardi, "New approaches for the repairs of memories with redundancy by row/column deletion for yield enhancement," IEEE Trans. Comput. Aided Des., vol. 9, pp. 323–328, Mar. 1990.
54. T. Kawagoe, J. Ohtani, M. Niiro, T. Ooishi, M. Hamada and H. Hidaka, "A built-in self-repair analyzer (CRESTA) for embedded DRAMs," in Proc. ITC, Oct. 2000, pp. 567–574.
55. C.-T. Huang, C.-F. Wu, J.-F. Li and C.-W. Wu, "Built-in redundancy analysis for memory yield improvement," IEEE Trans. Reliab., vol. 52, pp. 386–399, Dec. 2003.
56. N. Ohtsuka, S. Tanaka, J. Miyamoto, S. Saito, S. Atsumi, K. Imamiya, K. Yoshikawa, N. Matsukawa, S. Mori, N. Arai, T. Shinagawa, Y. Kaneko, J. Matsunaga and T. Iizuka, "A 4-Mbit CMOS EPROM," IEEE J. Solid-State Circuits, vol. SC-22, pp. 669–675, Oct. 1987.
57. D. Kantz, J. R. Goetz, R. Bender, M. Baehring, J. Wawersig, W. Meyer and W. Mueller, "A 256K DRAM with descrambled redundancy test capability," IEEE J. Solid-State Circuits, vol. SC-19, pp. 596–602, Oct. 1984.

Chapter 3
Error Checking and Correction (ECC)

3.1 Introduction

One of the key issues to design on-chip error checking and correction (ECC) circuits is the selection of an error-correcting code, which is the system to express data by a code, as described in Sect. 1.2. Error correcting codes proposed for the on-chip ECC of memory LSIs are *linear codes*. The linear codes feature that data are described by vectors and that coding (adding check bits) and decoding (checking and correcting errors) procedures are described by the operations between vectors and matrices. Therefore, the mathematical knowledge of *linear algebra* is required. In the case of an ordinary binary memory, a binary linear code is used, where each element of the vectors and matrices is “0” or “1.” In the case of a nonbinary (multilevel) memory, where plural bits are stored in a memory cell, a nonbinary linear code is more suitable, where each element may be not only “0” or “1” but also another value. What values should be used in addition to “0” and “1”? The answer is given by a set called *Galois field*.

At the design of on-chip ECC circuits, the chip-size and access-time penalties should be minimized, while maximizing error-correcting capability. However, there exists a trade-off between these parameters. Generally, a code with a higher error-correcting capability has a larger chip-size and access-time penalties. Therefore, a most suitable error-correcting code should be selected according to the requirements of each memory LSI, considering the reduction of error rates. Testing is another important issue for ECC design. It includes testing the extra memory cells for check bits, and testing the coding/decoding circuits.

In this chapter, the mathematical background of ECC is first described. The linear algebra and its application to the coding/decoding are presented in Sect. 3.2, while the Galois field is explained in Sect. 3.3. The various error correcting codes used for ECC and their coding/decoding circuits are described in Sects. 3.4 and 3.5, respectively. The effects of on-chip ECC, reduction of soft-error and hard-error rates, are estimated in Sect. 3.6. Section 3.7 presents some examples of on-chip ECC applications to memory LSIs and discusses various practical problems and solutions at the design of ECC circuits. Finally, the testing techniques for on-chip ECC are explained in Sect. 3.8.

M. Horiguchi and K. Itoh, *Nanoscale Memory Repair*, Integrated Circuits and Systems 1,
DOI 10.1007/978-1-4419-7958-2_3,

3.2 Linear Algebra and Linear Codes

In this section, linear algebra necessary for the design of on-chip ECC and its application to the coding/decoding procedures of linear codes are explained.

A set of k-bit data $d_0, d_1, \ldots, d_{k-1}$ is expressed by a k-dimensional column vector.[1]

$$\boldsymbol{d} = \begin{pmatrix} d_0 \\ d_1 \\ \vdots \\ d_{k-1} \end{pmatrix}. \tag{3.1}$$

The coding (adding check bits) and decoding (checking and correcting errors) procedures are described by operations between vectors and matrices as described below.

3.2.1 *Coding Procedure*

The coding procedure is performed by multiplying the vector $\boldsymbol{d}$ by an n-row, k-column ($n > k$) matrix (called *generator matrix*)[2]:

$$\boldsymbol{G} = \begin{pmatrix} g_{00} & g_{01} & \cdots & g_{0\,k-1} \\ g_{10} & g_{11} & \cdots & g_{1\,k-1} \\ \vdots & \vdots & \ddots & \vdots \\ g_{n-1\,0} & g_{n-1\,1} & \cdots & g_{n-1\,k-1} \end{pmatrix}. \tag{3.2}$$

That is, the n elements $w_0, w_1, \ldots, w_{n-1}$ of the n-dimensional column vector defined by

$$\boldsymbol{w} = \begin{pmatrix} w_0 \\ w_1 \\ \vdots \\ w_{n-1} \end{pmatrix} = \boldsymbol{Gd} \tag{3.3}$$

are the elements of the resulting code word. Here, the dimension of $\boldsymbol{d}$ must be equal to the number of columns of $\boldsymbol{G}$. The multiplication of (3.3) is calculated as

[1]A row vector $(d_0, d_1, \ldots, d_{k-1})$ is usually used. In this book, however, a column vector is used.
[2]Usually a generator matrix is defined as the transposed matrix (element g_{ij} and g_{ji} are replaced by each other) of $\boldsymbol{G}$ in this book.

Table 3.1 Addition and multiplication of modulo 2. These are also addition and multiplication on GF(2)

Addition of modulo 2			
$x + y$		x	
		0	1
y	0	0	1
	1	1	0
Multiplication of modulo 2			
$x \cdot y$		x	
		0	1
y	0	0	0
	1	0	1

$$w_i = g_{i0}d_0 + g_{i1}d_1 + \cdots + g_{i\,k-1}d_{k-1} \quad \text{mod } 2 \quad (i = 0, 1, \ldots, n-1). \tag{3.4}$$

Here, "mod 2" means calculation of modulo 2. If the calculation result is even, the final result is 0, and if the calculation result is odd, the final result is 1. The "modulo 2" addition and multiplication are shown in Table 3.1. They are the same as the ordinary addition and multiplication except for "1 + 1 = 0." The addition and multiplication are implemented by an exclusive OR gate and an AND gate, respectively.

For memory applications, it is favorable that the input data bits are written into memory cells as they are. This means that the first k bits of the code word $\boldsymbol{w}$ are identical to $\boldsymbol{d}$, which are followed by $(n - k)$ check bits. That is, $w_0 = d_0$, $w_1 = d_1$, …, $w_{k-1} = d_{k-1}$. In order to satisfy this condition, $g_{ii} = 1$ $(0 \leq i \leq k - 1)$, $g_{ij} = 0$ $(0 \leq i, j \leq k - 1, i \neq j)$. Thus, the upper k rows of $\boldsymbol{G}$ must be a $k \times k$ unit matrix[3] $\boldsymbol{E}_k$.

$$\boldsymbol{G} = \begin{pmatrix} \boldsymbol{E}_k \\ \hline \boldsymbol{P} \end{pmatrix} \tag{3.5}$$

$$\boldsymbol{E}_k = \begin{pmatrix} 1 & 0 & \cdots & 0 \\ 0 & 1 & \cdots & 0 \\ \vdots & \vdots & \ddots & \vdots \\ 0 & 0 & \cdots & 1 \end{pmatrix}. \tag{3.6}$$

[Example] Four bits of data 0, 1, 0, and 1 are described by a vector

$$\boldsymbol{d} = \begin{pmatrix} 0 \\ 1 \\ 0 \\ 1 \end{pmatrix}. \tag{3.7}$$

[3] A matrix with diagonal elements of "1" and the other elements of "0." For any vector $\boldsymbol{v}$, $\boldsymbol{E}_k \cdot \boldsymbol{v} = \boldsymbol{v}$, and for any matrix $\boldsymbol{M}$, $\boldsymbol{E}_k \cdot \boldsymbol{M} = \boldsymbol{M}$, $\boldsymbol{M} \cdot \boldsymbol{E}_k = \boldsymbol{M}$.

Coding by a generator matrix

$$G = \begin{pmatrix} 1 & 0 & 0 & 0 \\ 0 & 1 & 0 & 0 \\ 0 & 0 & 1 & 0 \\ 0 & 0 & 0 & 1 \\ 1 & 1 & 0 & 1 \\ 1 & 0 & 1 & 1 \\ 0 & 1 & 1 & 1 \end{pmatrix}, \tag{3.8}$$

results in

$$\boldsymbol{w} = \boldsymbol{Gd} = \begin{pmatrix} 0 \\ 1 \\ 0 \\ 1 \\ 0 \\ 1 \\ 0 \end{pmatrix}. \tag{3.9}$$

Thus, three check bits 0, 1, and 0 have been added after $\boldsymbol{d}$.

The code word $\boldsymbol{w}$ is written into memory. If no error occurs during write, retention, and read procedures, read data is equal to $\boldsymbol{w}$. However, the read data is different from $\boldsymbol{w}$ if some error occurs. The error is described by a vector

$$\boldsymbol{e} = \begin{pmatrix} e_0 \\ e_1 \\ \vdots \\ e_{n-1} \end{pmatrix}, \tag{3.10}$$

where $e_i = 0$ if no error occurs on w_i, otherwise $e_i = 1$. The read data is given by

$$\boldsymbol{r} = \begin{pmatrix} r_0 \\ r_1 \\ \vdots \\ r_{n-1} \end{pmatrix} = \boldsymbol{w} + \boldsymbol{e} = \begin{pmatrix} w_0 + e_0 \\ w_1 + e_1 \\ \vdots \\ w_{n-1} + e_{n-1} \end{pmatrix}. \tag{3.11}$$

Here, the addition is calculated with modulo 2.

3.2.2 Decoding Procedure

The decoding procedure is performed by multiplying the read data $\boldsymbol{r}$ by a matrix (called *check matrix*):

$$\boldsymbol{H} = \begin{pmatrix} h_{00} & h_{01} & \cdots & h_{0\,n-1} \\ h_{10} & h_{11} & \cdots & h_{1\,n-1} \\ \vdots & \vdots & \ddots & \vdots \\ h_{n-k-1\,0} & h_{n-k-1\,1} & \cdots & h_{n-k-1\,n-1} \end{pmatrix}. \tag{3.12}$$

The number of rows and columns of $\boldsymbol{H}$ is $(n - k)$ and n, respectively. Multiplying the read data $\boldsymbol{r}$ by $\boldsymbol{H}$ produces an $(n - k)$-dimensional vector $\boldsymbol{s}$ called syndrome:

$$\boldsymbol{s} = \begin{pmatrix} s_0 \\ s_1 \\ \vdots \\ s_{n-k-1} \end{pmatrix} = \boldsymbol{H}\boldsymbol{r} = \begin{pmatrix} h_{00} & h_{01} & \cdots & h_{0\,n-1} \\ h_{10} & h_{11} & \cdots & h_{1\,n-1} \\ \vdots & \vdots & \ddots & \vdots \\ h_{n-k-1\,0} & h_{n-k-1\,1} & \cdots & h_{n-k-1\,n-1} \end{pmatrix} \begin{pmatrix} r_0 \\ r_1 \\ \vdots \\ r_{n-1} \end{pmatrix}. \tag{3.13}$$

This is calculated by

$$s_i = h_{i0}r_0 + h_{i1}r_1 + \cdots + h_{i\,n-1}r_{n-1} \mod 2 \quad (i = 0, 1, \ldots, n - k - 1). \tag{3.14}$$

Here, $\boldsymbol{H}$ is given by

$$\boldsymbol{H} = \left(\boldsymbol{P} \;\vdots\; \boldsymbol{E}_{n-k} \right) \tag{3.15}$$

Matrix $\boldsymbol{H}$ is produced by arranging $\boldsymbol{P}$ (lower $(n - k)$ rows of $\boldsymbol{G}$) and an $(n - k) \times (n - k)$ unit matrix $\boldsymbol{E}_{n-k}$. If no error, $\boldsymbol{s} = 0$ ($s_0 = s_1 = \cdots = s_{n-k-1} = 0$) because

$$\boldsymbol{s} = \boldsymbol{H}\boldsymbol{r} = \boldsymbol{H}\boldsymbol{w} = \boldsymbol{H}\boldsymbol{G}\boldsymbol{d} = \left(\boldsymbol{P} \;\vdots\; \boldsymbol{E}_{n-k} \right) \begin{pmatrix} \boldsymbol{E}_k \\ \text{--------} \\ \boldsymbol{P} \end{pmatrix} \boldsymbol{d} \tag{3.16}$$

$$= (\boldsymbol{P}\boldsymbol{E}_k + \boldsymbol{E}_{n-k}\boldsymbol{P})\boldsymbol{d} = (\boldsymbol{P} + \boldsymbol{P})\boldsymbol{d} = \boldsymbol{0} \cdot \boldsymbol{d} = \boldsymbol{0}.$$

Note that $\boldsymbol{P} + \boldsymbol{P} = \boldsymbol{0}$ mod 2. An error is detected by a nonzero syndrome $\boldsymbol{s} \neq \boldsymbol{0}$. The erroneous bit(s) is presumed by analyzing $\boldsymbol{s}$ and the bit is corrected, which is described later.

[Example] If the read data is

$$\boldsymbol{r} = \begin{pmatrix} 0 \\ 1 \\ 0 \\ 1 \\ 0 \\ 1 \\ 0 \end{pmatrix},$$

multiplying by

$$\boldsymbol{H} = \begin{pmatrix} 1 & 1 & 0 & 1 & 1 & 0 & 0 \\ 1 & 0 & 1 & 1 & 0 & 1 & 0 \\ 0 & 1 & 1 & 1 & 0 & 0 & 1 \end{pmatrix}, \tag{3.17}$$

produces a syndrome

$$\boldsymbol{s} = \boldsymbol{H}\boldsymbol{r} = \begin{pmatrix} 1 & 1 & 0 & 1 & 1 & 0 & 0 \\ 1 & 0 & 1 & 1 & 0 & 1 & 0 \\ 0 & 1 & 1 & 1 & 0 & 0 & 1 \end{pmatrix} \begin{pmatrix} 0 \\ 1 \\ 0 \\ 1 \\ 0 \\ 1 \\ 0 \end{pmatrix} = \begin{pmatrix} 0 \\ 0 \\ 0 \end{pmatrix}. \tag{3.18}$$

Thus, we know that the read data has no error. If the read data is

$$\boldsymbol{r}' = \begin{pmatrix} 0 \\ 1 \\ 1 \\ 1 \\ 0 \\ 1 \\ 0 \end{pmatrix}$$

(r_2 is erroneous), multiplying $\boldsymbol{H}$ produces a syndrome

$$\boldsymbol{s}' = \boldsymbol{H}\boldsymbol{r}' = \begin{pmatrix} 1 & 1 & 0 & 1 & 1 & 0 & 0 \\ 1 & 0 & 1 & 1 & 0 & 1 & 0 \\ 0 & 1 & 1 & 1 & 0 & 0 & 1 \end{pmatrix} \begin{pmatrix} 0 \\ 1 \\ 1 \\ 1 \\ 0 \\ 1 \\ 0 \end{pmatrix} = \begin{pmatrix} 0 \\ 1 \\ 1 \end{pmatrix}. \tag{3.19}$$

Thus, we know that the read data has an error.

3.3 Galois Field

The discussion in the previous section assumed *binary* linear codes used for binary memories, where "modulo 2" calculations were used. However, there are nonbinary (multilevel) memories where plural bits are stored in a memory cell. *Nonbinary* linear codes are useful as error correcting codes for such memories as is described in Sect. 3.7. Since nonbinary codes utilize calculations on *Galois field*, instead of the "modulo 2" calculations, the concept of Galois field is explained in this section.

A *field* is a set in which addition, subtraction, multiplication, and division are defined except for division by zero. For example, the set $\boldsymbol{R}$ composed of all real numbers is a field. The set $\boldsymbol{Q}$ composed of all rational numbers is also a field. However, the set $\boldsymbol{Z}$ composed of all integers is not a field because the quotient of an integer and another integer is not always an integer. Although $\boldsymbol{R}$ and $\boldsymbol{Q}$ have innumerable elements, there are fields having finite elements. A field composed of finite elements is called a finite field or a *Galois field*. It is known that the number of elements of a Galois field is equal to a prime number or a power of a prime number. Therefore, a Galois field with six or ten elements does not exist. A Galois field with q elements is described as GF(q). Since important cases for memory applications are q is equal to 2 or a power of 2, the discussion below is focused on these cases.

GF(2) is the "modulo 2" calculations described above. If we define the addition and multiplication of a set composed of "0" and "1" as shown in Table 3.1, the set {0, 1} becomes a field. However, GF(4), GF(8), etc. are not so simple. For example, "modulo 4" calculations among "0," "1," "2," and "3" are shown in Table 3.2. The set {0, 1, 2, 3}, however, is not a field because $1 \div 2$ is not defined (x that satisfies $2x = 1$ does not exist in this set). How to construct GF(2^m) based on GF(2) is explained below.

Table 3.2 Addition and multiplication of modulo 4. Set {0, 1, 2, 3} does not form a Galois field

Addition of modulo 4					
$x + y$		x			
		0	1	2	3
y	0	0	1	2	3
	1	1	2	3	0
	2	2	3	0	1
	3	3	0	1	2
Multiplication of modulo 4					
$x \cdot y$		x			
		0	1	2	3
y	0	0	0	0	0
	1	0	1	2	3
	2	0	2	0	2
	3	0	3	2	1

Table 3.3 Irreducible polynomials on GF(2)

m	m-th order polynomial
2	$x^2 + x + 1$
3	$x^3 + x + 1$
4	$x^4 + x + 1$
5	$x^5 + x^2 + 1$
6	$x^6 + x + 1$
7	$x^7 + x + 1$
8	$x^8 + x^4 + x^3 + x^2 + 1$

First, an irreducible polynomial on GF(2) with degree m:

$$f(x) = a_m x^m + a_{m-1} x^{m-1} + \cdots + a_1 x + a_0 \tag{3.20}$$

is selected. "Irreducible" means that the polynomial cannot be factorized on GF(2). Note that an irreducible polynomial of ordinary meaning (on $\boldsymbol{R}$) is not always irreducible on GF(2). Polynomial $x^2 + 1$ is irreducible on $\boldsymbol{R}$, but is not irreducible on GF(2) because it is factorized as

$$x^2 + 1 = x^2 + 2x + 1 = (x + 1)^2 \tag{3.21}$$

(note that $2 = 0 \bmod 2$). On the other hand, polynomial $x^2 + x + 1$ is irreducible not only on $\boldsymbol{R}$ but also on GF(2). Examples of irreducible polynomials with degrees 2–8 are shown in Table 3.3. If we define α as a root of the equation $f(x) = 0$:

$$a_m \alpha^m + a_{m-1} \alpha^{m-1} + \cdots + a_1 \alpha + a_0 = 0, \tag{3.22}$$

a set composed of $\alpha^0(=1)$, α, α^2, α^3,…, α^{2m-2}, and zero is a field.

[Example] A quadratic irreducible polynomial $x^2 + x + 1$ is selected to construct GF(4). Let us define α as a root of equation $x^2 + x + 1 = 0$ and consider the set composed of $1, \alpha, \alpha^2$, and zero. The addition and multiplication between two elements of the set are defined as follows. Since $\alpha^2 + \alpha + 1 = 0$, $\alpha^2 = -\alpha - 1 = \alpha + 1$. Therefore, the additions are defined as $\alpha + 1 = \alpha^2$, $\alpha^2 + 1 = (\alpha + 1) + 1 = \alpha$, $\alpha^2 + \alpha = (\alpha + 1) + \alpha = 1, 1 + 1 = 2 = 0, \alpha + \alpha = 2\alpha = 0$, and $\alpha^2 + \alpha^2 = 2\alpha^2 = 0$, as shown in Table 3.4. The multiplications between the elements are calculated noting that $\alpha^3 = \alpha^2\alpha = (\alpha + 1)\alpha = \alpha^2 + \alpha = (\alpha + 1) + \alpha = 1$ as shown in Table 3.4. These tables mean that the set $\{1, \alpha, \alpha^2, 0\}$ is a field.

In similar ways GF(8) is constructed using a cubic irreducible polynomial $x^3 + x + 1$, and GF(16) is constructed using a quartic irreducible polynomial $x^4 + x + 1$. The additions and multiplications between two elements of GF(8) and GF(16) are summarized in Tables 3.5 and 3.6, respectively.

Table 3.4 Addition and multiplication on GF(4). α is a root of $x^2 + x + 1 = 0$

Addition on GF(4)

$x + y$		x			
		0	1	α	α^2
y	0	0	1	α	α^2
	1	1	0	α^2	α
	α	α	α^2	0	1
	α^2	α^2	α	1	0

Multiplication on GF(4)

$x \cdot y$		x			
		0	1	α	α^2
y	0	0	0	0	0
	1	0	1	α	α^2
	α	0	α	α^2	1
	α^2	0	α^2	1	α

Table 3.5 Addition and multiplication on GF(8). β is a root of $x^3 + x + 1 = 0$

Addition on GF(8)

$x + y$		x							
		0	1	β	β^2	β^3	β^4	β^5	β^6
y	0	0	1	β	β^2	β^3	β^4	β^5	β^6
	1	1	0	β^3	β^6	β	β^5	β^4	β^2
	β	β	β^3	0	β^4	1	β^2	β^6	β^5
	β^2	β^2	β^6	β^4	0	β^5	β	β^3	1
	β^3	β^3	β	1	β^5	0	β^6	β^2	β^4
	β^4	β^4	β^5	β^2	β	β^6	0	1	β^3
	β^5	β^5	β^4	β^6	β^3	β^2	1	0	β
	β^6	β^6	β^2	β^5	1	β^4	β^3	β	0

Multiplication on GF(8)

$x \cdot y$		x							
		0	1	β	β^2	β^3	β^4	β^5	β^6
y	0	0	0	0	0	0	0	0	0
	1	0	1	β	β^2	β^3	β^4	β^5	β^6
	β	0	β	β^2	β^3	β^4	β^5	β^6	1
	β^2	0	β^2	β^3	β^4	β^5	β^6	1	β
	β^3	0	β^3	β^4	β^5	β^6	1	β	β^2
	β^4	0	β^4	β^5	β^6	1	β	β^2	β^3
	β^5	0	β^5	β^6	1	β	β^2	β^3	β^4
	β^6	0	β^6	1	β	β^2	β^3	β^4	β^5

3.4 Error-Correcting Codes

In this section, representative error-correcting codes proposed for the on-chip ECC of memory LSIs are described. For more details of error-correcting codes, refer the textbooks on code theory [1].

Table 3.6 Addition and multiplication on GF(16). γ is a root of $x^4 + x + 1 = 0$

Addition on GF(16)

$x + y$		x															
		0	1	γ	γ^2	γ^3	γ^4	γ^5	γ^6	γ^7	γ^8	γ^9	γ^{10}	γ^{11}	γ^{12}	γ^{13}	γ^{14}
y	0	0	1	γ	γ^2	γ^3	γ^4	γ^5	γ^6	γ^7	γ^8	γ^9	γ^{10}	γ^{11}	γ^{12}	γ^{13}	γ^{14}
	1	1	0	γ^4	γ^8	γ^{14}	γ	γ^{10}	γ^{13}	γ^9	γ^2	γ^7	γ^5	γ^{12}	γ^{11}	γ^6	γ^3
	γ	γ	γ^4	0	γ^5	γ^9	1	γ^2	γ^{11}	γ^{14}	γ^{10}	γ^3	γ^8	γ^6	γ^{13}	γ^{12}	γ^7
	γ^2	γ^2	γ^8	γ^5	0	γ^6	γ^{10}	γ	γ^3	γ^{12}	1	γ^{11}	γ^4	γ^9	γ^7	γ^{14}	γ^{13}
	γ^3	γ^3	γ^{14}	γ^9	γ^6	0	γ^7	γ^{11}	γ^2	γ^4	γ^{13}	γ	γ^{12}	γ^5	γ^{10}	γ^8	1
	γ^4	γ^4	γ	1	γ^{10}	γ^7	0	γ^8	γ^{12}	γ^3	γ^5	γ^{14}	γ^2	γ^{13}	γ^6	γ^{11}	γ^9
	γ^5	γ^5	γ^{10}	γ^2	γ	γ^{11}	γ^8	0	γ^9	γ^{13}	γ^4	γ^6	1	γ^3	γ^{14}	γ^7	γ^{12}
	γ^6	γ^6	γ^{13}	γ^{11}	γ^3	γ^2	γ^{12}	γ^9	0	γ^{10}	γ^{14}	γ^5	γ^7	γ	γ^4	1	γ^8
	γ^7	γ^7	γ^9	γ^{14}	γ^{12}	γ^4	γ^3	γ^{13}	γ^{10}	0	γ^{11}	1	γ^6	γ^8	γ^2	γ^5	γ
	γ^8	γ^8	γ^2	γ^{10}	1	γ^{13}	γ^5	γ^4	γ^{14}	γ^{11}	0	γ^{12}	γ	γ^7	γ^9	γ^3	γ^6
	γ^9	γ^9	γ^7	γ^3	γ^{11}	γ	γ^{14}	γ^6	γ^5	1	γ^{12}	0	γ^{13}	γ^2	γ^8	γ^{10}	γ^4
	γ^{10}	γ^{10}	γ^5	γ^8	γ^4	γ^{12}	γ^2	1	γ^7	γ^6	γ	γ^{13}	0	γ^{14}	γ^3	γ^9	γ^{11}
	γ^{11}	γ^{11}	γ^{12}	γ^6	γ^9	γ^5	γ^{13}	γ^3	γ	γ^8	γ^7	γ^2	γ^{14}	0	1	γ^4	γ^{10}
	γ^{12}	γ^{12}	γ^{11}	γ^{13}	γ^7	γ^{10}	γ^6	γ^{14}	γ^4	γ^2	γ^9	γ^8	γ^3	1	0	γ	γ^5
	γ^{13}	γ^{13}	γ^6	γ^{12}	γ^{14}	γ^8	γ^{11}	γ^7	1	γ^5	γ^3	γ^{10}	γ^9	γ^4	γ	0	γ^2
	γ^{14}	γ^{14}	γ^3	γ^7	γ^{13}	1	γ^9	γ^{12}	γ^8	γ	γ^6	γ^4	γ^{11}	γ^{10}	γ^5	γ^2	0

Multiplication on GF(16)

$x \cdot y$		x															
		0	1	γ	γ^2	γ^3	γ^4	γ^5	γ^6	γ^7	γ^8	γ^9	γ^{10}	γ^{11}	γ^{12}	γ^{13}	γ^{14}
y	0	0	0	0	0	0	0	0	0	0	0	0	0	0	0	0	0
	1	0	1	γ	γ^2	γ^3	γ^4	γ^5	γ^6	γ^7	γ^8	γ^9	γ^{10}	γ^{11}	γ^{12}	γ^{13}	γ^{14}
	γ	0	γ	γ^2	γ^3	γ^4	γ^5	γ^6	γ^7	γ^8	γ^9	γ^{10}	γ^{11}	γ^{12}	γ^{13}	γ^{14}	1
	γ^2	0	γ^2	γ^3	γ^4	γ^5	γ^6	γ^7	γ^8	γ^9	γ^{10}	γ^{11}	γ^{12}	γ^{13}	γ^{14}	1	γ
	γ^3	0	γ^3	γ^4	γ^5	γ^6	γ^7	γ^8	γ^9	γ^{10}	γ^{11}	γ^{12}	γ^{13}	γ^{14}	1	γ	γ^2
	γ^4	0	γ^4	γ^5	γ^6	γ^7	γ^8	γ^9	γ^{10}	γ^{11}	γ^{12}	γ^{13}	γ^{14}	1	γ	γ^2	γ^3
	γ^5	0	γ^5	γ^6	γ^7	γ^8	γ^9	γ^{10}	γ^{11}	γ^{12}	γ^{13}	γ^{14}	1	γ	γ^2	γ^3	γ^4
	γ^6	0	γ^6	γ^7	γ^8	γ^9	γ^{10}	γ^{11}	γ^{12}	γ^{13}	γ^{14}	1	γ	γ^2	γ^3	γ^4	γ^5
	γ^7	0	γ^7	γ^8	γ^9	γ^{10}	γ^{11}	γ^{12}	γ^{13}	γ^{14}	1	γ	γ^2	γ^3	γ^4	γ^5	γ^6
	γ^8	0	γ^8	γ^9	γ^{10}	γ^{11}	γ^{12}	γ^{13}	γ^{14}	1	γ	γ^2	γ^3	γ^4	γ^5	γ^6	γ^7
	γ^9	0	γ^9	γ^{10}	γ^{11}	γ^{12}	γ^{13}	γ^{14}	1	γ	γ^2	γ^3	γ^4	γ^5	γ^6	γ^7	γ^8
	γ^{10}	0	γ^{10}	γ^{11}	γ^{12}	γ^{13}	γ^{14}	1	γ	γ^2	γ^3	γ^4	γ^5	γ^6	γ^7	γ^8	γ^9
	γ^{11}	0	γ^{11}	γ^{12}	γ^{13}	γ^{14}	1	γ	γ^2	γ^3	γ^4	γ^5	γ^6	γ^7	γ^8	γ^9	γ^{10}
	γ^{12}	0	γ^{12}	γ^{13}	γ^{14}	1	γ	γ^2	γ^3	γ^4	γ^5	γ^6	γ^7	γ^8	γ^9	γ^{10}	γ^{11}
	γ^{13}	0	γ^{13}	γ^{14}	1	γ	γ^2	γ^3	γ^4	γ^5	γ^6	γ^7	γ^8	γ^9	γ^{10}	γ^{11}	γ^{12}
	γ^{14}	0	γ^{14}	1	γ	γ^2	γ^3	γ^4	γ^5	γ^6	γ^7	γ^8	γ^9	γ^{10}	γ^{11}	γ^{12}	γ^{13}

3.4.1 Minimum Distance

Minimum distance is an important concept for discussing ECC. It is the minimum value of *Hamming distance* $d(\boldsymbol{v}_1, \boldsymbol{v}_2)$ between two vectors $\boldsymbol{v}_1$ and $\boldsymbol{v}_2$ defined as the number of different bits when compared bit-by-bit. If,

$$\boldsymbol{v}_1 = \begin{pmatrix} 0 \\ 1 \\ 0 \\ 1 \\ 0 \end{pmatrix}$$

and

$$\boldsymbol{v}_2 = \begin{pmatrix} 0 \\ 1 \\ 1 \\ 0 \\ 0 \end{pmatrix}, d(\boldsymbol{v}_1, \boldsymbol{v}_2) = 2$$

because the third and fourth bits are different. If t errors occur in a code word $\boldsymbol{w}$ written into a memory and $\boldsymbol{r}$ is read out, $d(\boldsymbol{w}, \boldsymbol{r}) = t$. Thus, Hamming distance corresponds to the number of errors. The minimum distance of a code $d_{\min}$ is defined as the minimum Hamming distance between two code words of the code.

$$d_{\min} = \min_{i \neq j} d(\boldsymbol{w}_i, \boldsymbol{w}_j). \tag{3.23}$$

If $d_{\min} = 2$, we can detect a single-bit error. Because the vectors $\boldsymbol{v}_1$, $\boldsymbol{v}_2$, $\boldsymbol{v}_3$, etc., which are placed at Hamming distance of one from code word $\boldsymbol{w}_1$, are not code words. However, we cannot correct the error. If, for example, $\boldsymbol{v}_3$ is read out, we cannot judge whether an error has occurred at the third bit of $\boldsymbol{w}_1$ or at the fourth bit of $\boldsymbol{w}_2$, as shown in Fig. 3.1a. If $d_{\min} = 3$, we can correct a single-bit error. If $\boldsymbol{v}_1$ is read out, the write data is assumed as $\boldsymbol{w}_1$, and if $\boldsymbol{v}_2$ was read out, the write data is assumed as $\boldsymbol{w}_2$ as shown in Fig. 3.1b. Generally, if $d_{\min} = 2t + 1$, we can correct t errors or fewer. If $d_{\min} = 2t + 2$, we can correct t errors or fewer and can detect $(t + 1)$ errors.

3.4.2 Number of Check Bits

Fewer number of check bits is desired to reduce the chip-area penalty due to the on-chip ECC. However, there exists a lower bound of check-bit count to realize a certain error detecting/correcting capability. The bound is calculated as follows.

First, let us consider the case of single-error correction (SEC) of adding check bits to k-bit data and making an n-bit code word. The number of cases of a single-bit error occurring on a code word $\boldsymbol{w}_1$ is n, as indicated by the circle C_1 in Fig. 3.1b (note the possibility of an error occurring at one of the check bits). Therefore, the number of vectors in C_1 (Hamming distance from $\boldsymbol{w}_1$ is within one) is $n + 1$, including no-error case. Similarly, we can draw another circle C_2 around code

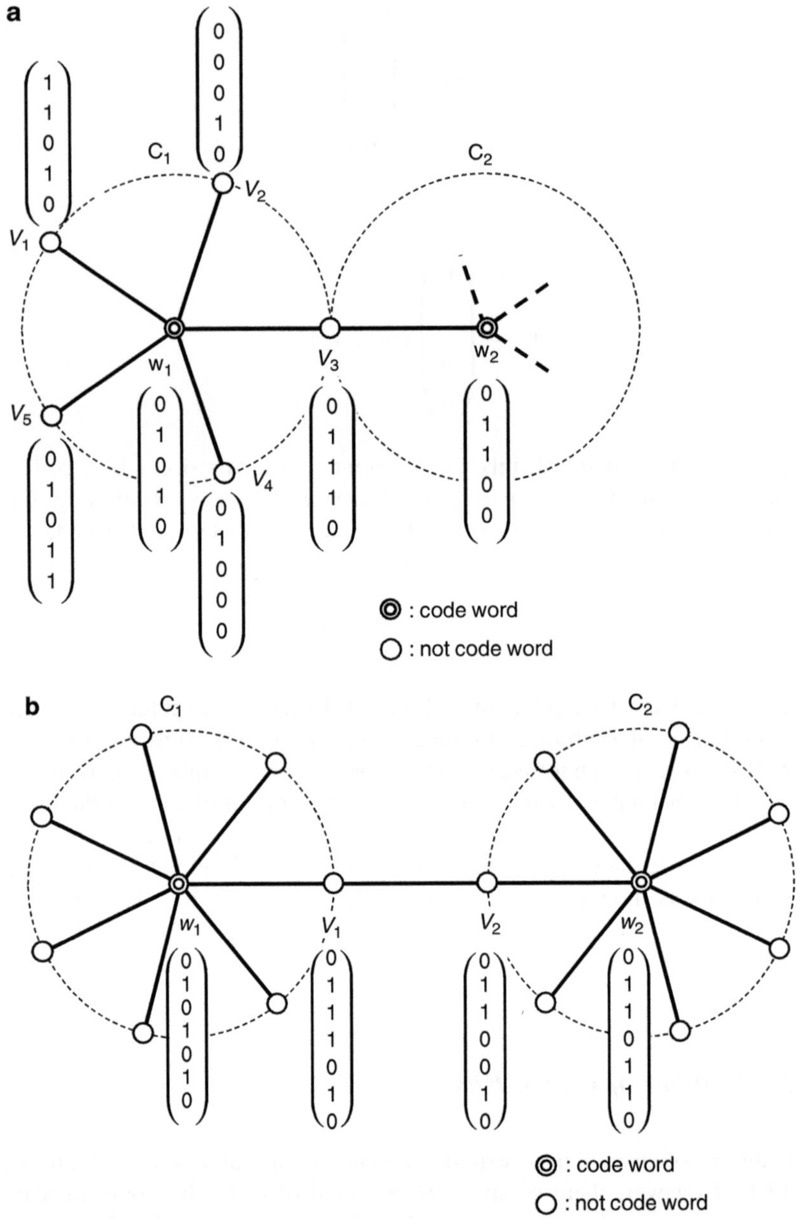

Fig. 3.1 Minimum distance and error detecting/correcting capability: (**a**) case of $d_{\min} = 2$ and (**b**) case of $d_{\min} = 3$

word $\boldsymbol{w}_2$. Two circles C_1 and C_2 must be exclusive. If there were a vector $\boldsymbol{v}_3$ common to C_1 and C_2 as shown in Fig. 3.1a, we could not judge whether the write data were $\boldsymbol{w}_1$ or $\boldsymbol{w}_2$ from the read data $\boldsymbol{v}_3$. Since the number of code words is

Table 3.7 Lower bound of the number of required check bits for single-error correction (SEC), single-error correction and double-error detection (SEC–DEC), and double-error correction (DEC) codes

Data-bit count, k	4	8	16	32	64	128	256
SEC	3	4	5	6	7	8	9
SEC–DED	4	5	6	7	8	9	10
DEC	6	7	9	10	12	14	16

Note that these numbers are lower bounds. A code with a check-bit count shown here does not necessarily exist

2^k, the total number of vectors in each circle around each code word is $2^k(n + 1)$. On the other hand, the number of n-dimensional vectors is 2^n. Thus, the following inequality must be satisfied.

$$2^k(n+1)\leqslant 2^n, \quad 2^{n-k}\geqslant n+1. \tag{3.24}$$

From another viewpoint, the relationship (3.24) means that we must discriminate $(n + 1)$ cases using $(n - k)$-bit syndrome.

Next, let us consider the case of double-error correction (DEC). In this case, each circle includes (1) 1 vector of no-error case, (2) n vectors of single-error cases, and (3) $\binom{n}{2} = n(n-1)/2$ cases of double-error cases. Therefore, the inequality (3.24) is replaced by

$$2^{n-k}\geqslant 1+n+\frac{n(n-1)}{2}. \tag{3.25}$$

Generally for t-bit error correction the following relationship must be satisfied.

$$2^{n-k}\geqslant \sum_{i=0}^{t}\binom{n}{i}. \tag{3.26}$$

For t-bit error correction and $(t + 1)$-bit error detection, the number of required check bits is greater than that of t-bit error correction by one bit.

Table 3.7 summarizes the required check bits for SEC, SEC–DEC, and DEC. The ratio of required check bits $(n - k)/k$ decreases with the increase of data-bit count k in each case. Note that these numbers are theoretical lower bounds calculated from (3.26). A code with a check-bit count shown in the table does not necessarily exist. As for SEC and SEC–DED, however, the lower bounds are realized by Hamming code and extended Hamming code, respectively, as described later.

3.4.3 Single Parity Check Code

A check bit (parity bit) p is added to k-bit data

$$\boldsymbol{d} = \begin{pmatrix} d_0 \\ d_1 \\ \vdots \\ d_{k-1} \end{pmatrix}$$

producing a code word

$$\boldsymbol{w} = \begin{pmatrix} d_0 \\ d_1 \\ \vdots \\ d_{k-1} \\ p \end{pmatrix},$$

so that the number of "1"s in $\boldsymbol{w}$ is even. The minimum distance of this code is two. We can, therefore, detect a single-bit error, but cannot correct it. The generator matrix and the check matrix are expressed as

$$\boldsymbol{G} = \begin{pmatrix} 1 & 0 & \cdots & 0 \\ 0 & 1 & \cdots & 0 \\ \vdots & \vdots & \ddots & \vdots \\ 0 & 0 & \cdots & 1 \\ 1 & 1 & \cdots & 1 \end{pmatrix}, \tag{3.27}$$

and

$$\boldsymbol{H} = (\,1 \quad 1 \quad \cdots \quad 1 \quad 1\,), \tag{3.28}$$

respectively. The generation of the parity bit is calculated by

$$p = d_0 + d_1 + \cdots + d_{k-1} \quad \text{mod } 2. \tag{3.29}$$

The error detection is calculated by

$$s = d_0 + d_1 + \cdots + d_{k-1} + p \quad \text{mod } 2. \tag{3.30}$$

3.4.4 Hamming Code

A Hamming code is defined as the code with a check matrix in which all the column vectors except for zero vector are arranged. The minimum distance is three and we can correct a single-bit error. The number of rows of the check matrix is $n - k$, and

Table 3.8 Relationship among code length, number of check bits, and number of data bits of Hamming code and shortened Hamming code

Hamming code	Code length, n	7	15	31	63	127	255
	Data-bit count, k	4	11	26	57	120	247
Shortened Hamming code	Code length, n	–	12	21	38	71	136
	Data-bit count, k	–	8	16	32	64	128
Check-bit count, $n - k$		3	4	5	6	7	8

the number of columns is equal to the code length n. Since the number of $(n-k)$-dimensional vectors except for zero vector is $2^{n-k} - 1$, the following equation is satisfied.

$$2^{n-k} - 1 = n. \tag{3.31}$$

For example, (3.17) is a check matrix of a Hamming code ($n = 7$, $k = 4$). Comparing (3.31) and (3.24), we find that the lower bound of check-bit count is realized. Thus, Hamming codes are the most efficient (with fewest check-bit counts) single-error correcting codes. The relationship between code length, check-bit count, and data-bit count is shown in Table 3.8. The ratio of check bits decreases with the increase of data-bit count.

[Example]

$$\boldsymbol{H}_1 = \begin{pmatrix} 1 & 1 & 1 & 0 & 0 & 0 & 1 & 1 & 1 & 0 & 1 & 1 & 0 & 0 & 0 \\ 1 & 0 & 0 & 1 & 1 & 0 & 1 & 1 & 0 & 1 & 1 & 0 & 1 & 0 & 0 \\ 0 & 1 & 0 & 1 & 0 & 1 & 1 & 0 & 1 & 1 & 1 & 0 & 0 & 1 & 0 \\ 0 & 0 & 1 & 0 & 1 & 1 & 0 & 1 & 1 & 1 & 1 & 0 & 0 & 0 & 1 \end{pmatrix} \tag{3.32}$$

is the check matrix of the Hamming code with a code length of 15, data-bit count of 11, and check-bit count of 4. The right four columns of the matrix must be an 4×4 unit matrix, and the left eleven columns are composed of column vectors with two or more "1"s.

For memory applications the data-bit count k is usually a power of 2. A shortened Hamming code is used in such cases. "Shortened" means that some of k data bits are not used (assumed to be zero). Table 3.8 also shows shortened Hamming codes with data-bit counts of powers of 2. If data-bit count is 2^m, the required check-bit count is $m + 1$.

[Example] Let us consider shortening the Hamming code given in (3.32) so that data-bit count is eight. The criteria of the shortening are as follows.

1. The number of "1"s in each row of the check matrix should be minimized because it corresponds to the number of exclusive OR gates in the decoding circuit.
2. The number of "1"s in each row should be as uniform as possible. Since the time for calculating each syndrome bit corresponds to the number of "1"s of each row, the delay time of the decoding circuit is determined by the row with maximum "1" count.

Deleting the 9–11th columns of $\boldsymbol{H}_1$ according to these criteria, we obtain

$$\boldsymbol{H}_2 = \begin{pmatrix} 1 & 1 & 1 & 0 & 0 & 0 & 1 & 1 & 1 & 0 & 0 & 0 \\ 1 & 0 & 0 & 1 & 1 & 0 & 1 & 1 & 0 & 1 & 0 & 0 \\ 0 & 1 & 0 & 1 & 0 & 1 & 1 & 0 & 0 & 0 & 1 & 0 \\ 0 & 0 & 1 & 0 & 1 & 1 & 0 & 1 & 0 & 0 & 0 & 1 \end{pmatrix}. \tag{3.33}$$

Hamming codes feature that the ratio of required check bits $(n - k)/k$ is low and that the coding/decoding procedures are relatively simple. Therefore, they are widely used for on-chip error correction of memory LSIs as described in Sect. 3.7.

3.4.5 Extended Hamming Code and Hsiao Code

Adding a parity bit to a Hamming code (or shortened Hamming code) just like the single-parity-check code results in an extended Hamming code. The minimum distance is four and we can correct a single error and detect a double error. The check matrix is given by

$$\boldsymbol{H} = \left(\begin{array}{ccc|c} & & & 0 \\ & \boldsymbol{H}_0 & & \vdots \\ & & & 0 \\ \hline 1 & \cdots & 1 & 1 \end{array}\right) \tag{3.34}$$

where $\boldsymbol{H}_0$ is the check matrix of the original Hamming code.

[Example] From the shortened Hamming code of (3.33), an extended Hamming code

$$\boldsymbol{H}_3 = \begin{pmatrix} 1 & 1 & 1 & 0 & 0 & 0 & 1 & 1 & 1 & 0 & 0 & 0 & 0 \\ 1 & 0 & 0 & 1 & 1 & 0 & 1 & 1 & 0 & 1 & 0 & 0 & 0 \\ 0 & 1 & 0 & 1 & 0 & 1 & 1 & 0 & 0 & 0 & 1 & 0 & 0 \\ 0 & 0 & 1 & 0 & 1 & 1 & 0 & 1 & 0 & 0 & 0 & 1 & 0 \\ 1 & 1 & 1 & 1 & 1 & 1 & 1 & 1 & 1 & 1 & 1 & 1 & 1 \end{pmatrix} \tag{3.35}$$

is obtained.

A variation of the extended Hamming code was proposed by Hsiao [2]. A row is added to the check matrix of the original Hamming code so that the number of "1"s of each column is odd.

[Example] Deleting the 1st, 6th, and 11th columns of the check matrix of (3.32) results in the check matrix of a shortened Hamming code:

$$\boldsymbol{H}_4 = \begin{pmatrix} 1 & 1 & 0 & 0 & 1 & 1 & 1 & 0 & 1 & 0 & 0 & 0 \\ 0 & 0 & 1 & 1 & 1 & 1 & 0 & 1 & 0 & 1 & 0 & 0 \\ 1 & 0 & 1 & 0 & 1 & 0 & 1 & 1 & 0 & 0 & 1 & 0 \\ 0 & 1 & 0 & 1 & 0 & 1 & 1 & 1 & 0 & 0 & 0 & 1 \end{pmatrix}. \tag{3.36}$$

From this, the matrix of a Hsiao code:

$$\boldsymbol{H}_5 = \begin{pmatrix} 1 & 1 & 0 & 0 & 1 & 1 & 1 & 0 & 1 & 0 & 0 & 0 & 0 \\ 0 & 0 & 1 & 1 & 1 & 1 & 0 & 1 & 0 & 1 & 0 & 0 & 0 \\ 1 & 0 & 1 & 0 & 1 & 0 & 1 & 1 & 0 & 0 & 1 & 0 & 0 \\ 0 & 1 & 0 & 1 & 0 & 1 & 1 & 1 & 0 & 0 & 0 & 1 & 0 \\ 1 & 1 & 1 & 1 & 0 & 0 & 0 & 0 & 0 & 0 & 0 & 0 & 1 \end{pmatrix} \quad (3.37)$$

is obtained. Here, we used (3.36) instead of (3.33) because the number of "1"s in each row of the final resultant matrix (3.37) should be as uniform as possible (criterion 2 above). This code features that the check matrix has fewer "1"s than (3.35) and that the number of "1"s in each row is more uniform. The hardware size and delay time of the decoding circuit are therefore reduced.

3.4.6 Bidirectional Parity Code

To construct a bidirectional parity code, $k_x \times k_y$ data bits are arranged to form a rectangle as shown in Fig. 3.2. Next, check bits are added so that each row/column contains even number of "1"s. The check bit count is $(k_x + k_y + 1)$. The minimum distance is four and we can correct a single error and detect a double error. The check matrix is shown in Fig. 3.3. Although this code requires more check bits than the Hamming code, the coding/decoding procedures are simpler. So the application to on-chip ECC of memory LSIs is proposed as described in Sect. 3.7.

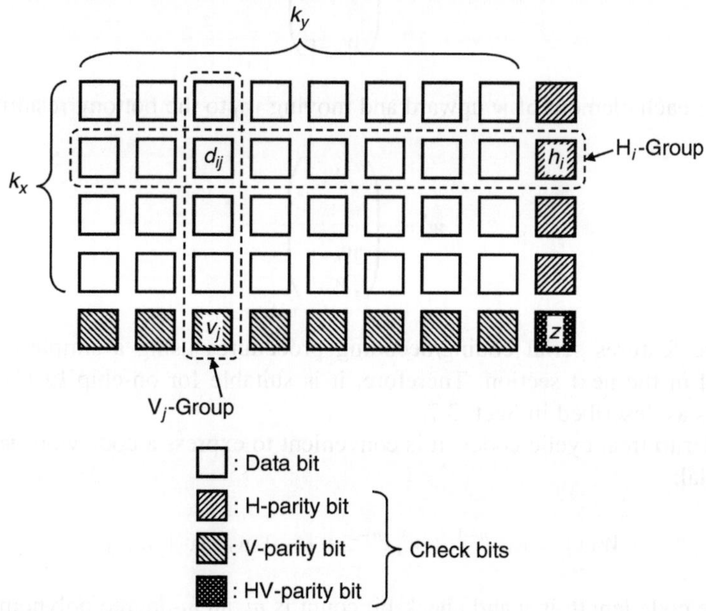

Fig. 3.2 Bidirectional parity code

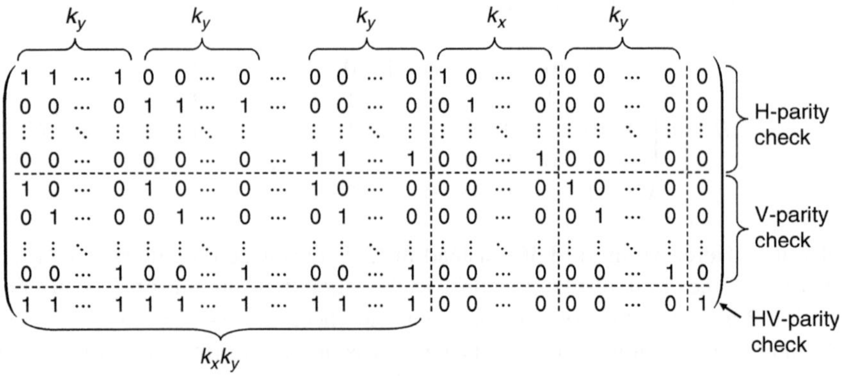

Fig. 3.3 Check matrix of bidirectional parity code

3.4.7 *Cyclic Code*

Cyclic code is a subset of linear code. It features that a *cyclic replacement* of a code word is also a code word. The cyclic replacement of a code word

$$\boldsymbol{w} = \begin{pmatrix} w_0 \\ w_1 \\ \vdots \\ w_{n-1} \end{pmatrix}$$

is shifting each element of $\boldsymbol{w}$ upward and moving w_0 to the bottom, resulting in

$$\boldsymbol{w}' = \begin{pmatrix} w_1 \\ \vdots \\ w_{n-1} \\ w_0 \end{pmatrix}.$$

This code features serial coding/decoding procedures using a simple circuit as discussed in the next section. Therefore, it is suitable for on-chip ECC of serial memories as described in Sect. 3.7.

In order to treat cyclic codes, it is convenient to express a code word $\boldsymbol{w}$ using a polynomial:

$$W(x) = w_0 x^{n-1} + w_1 x^{n-2} + \cdots + w_{n-2} x + w_{n-1}. \tag{3.38}$$

When the code length is n and check-bit count is m, an m-degree polynomial $G(x)$ (generator polynomial) that can divide $x^n + 1$ is selected as

$$G(x) = x^m + g_1x^{m-1} + \cdots + g_{m-1}x + g_m. \qquad (3.39)$$

If we construct a code composed of the polynomials that can be divided by $G(x)$, the code is a cyclic code. The cyclic replacement of $\boldsymbol{w}$, $\boldsymbol{w}'$, is expressed by the polynomial:

$$\begin{aligned} W'(x) &= w_1x^{n-1} + w_2x^{n-2} + \cdots + w_{n-1}x + w_0 \\ &= x(w_0x^{n-1} + w_1x^{n-2} + \cdots + w_{n-2}x + w_{n-1}) + w_0(1 - x^n) \qquad (3.40) \\ &= xW(x) + w_0(x^n + 1) \end{aligned}$$

(note the "modulo 2" calculation; $-1 = 1 \bmod 2$). If $W(x)$ can be divided by $G(x)$ ($\boldsymbol{w}$ is a code word), $W'(x)$ can also be divided by $G(x)$ ($\boldsymbol{w}'$ is a code word). The coding procedure is as follows. Data bits $d_0, d_1, \ldots, d_{k-1}$ are expressed by a polynomial:

$$D(x) = d_0x^{k-1} + d_1x^{k-2} + \cdots + d_{k-2}x + d_{k-1}. \qquad (3.41)$$

Multiplying this polynomial by x^m and dividing by $G(x)$ produces a residue polynomial $C(x)$ expressed as

$$D(x) \cdot x^m = G(x)Q(x) + C(x), \qquad (3.42)$$

$$C(x) = c_0x^{m-1} + c_1x^{m-2} + \cdots + c_{m-2}x + c_{m-1}, \qquad (3.43)$$

where $Q(x)$ is a polynomial. The check bits are the coefficients of $C(x)$. The code word is therefore given by

$$\begin{aligned} W(x) &= D(x) \cdot x^m + C(x) \\ &= d_0x^{n-1} + d_1x^{n-2} + \cdots + d_{k-1}x^m + c_0x^{m-1} \qquad (3.44) \\ &+ c_1x^{m-2} + \cdots + c_{m-2}x + c_{m-1}. \end{aligned}$$

It is clear from (3.42) and (3.44) that $W(x)$ can be divided by $G(x)$. The decoding procedure is as follows. The read data is expressed by a polynomial

$$R(x) = r_0x^{n-1} + r_1x^{n-2} + \cdots + r_{n-2}x + r_{n-1}. \qquad (3.45)$$

Dividing by $G(x)$ generates a residue, which is the syndrome $S(x)$:

$$R(x) = G(x)Q'(x) + S(x), \qquad (3.46)$$

$$S(x) = s_0x^{m-1} + s_1x^{m-2} + \cdots + s_{m-2}x + s_{m-1}, \qquad (3.47)$$

where $Q'(x)$ is a polynomial. It is clear from the above discussion that $S(x) = 0$ if no error.

[Example] Polynomial of degree $7x^7 + 1$ can be divided by a cubic polynomial $G(x) = x^3 + x + 1$ because $x^7 + 1 = (x^3 + x + 1)(x^4 + x^2 + x + 1)$. Let us construct a cyclic code with a generator polynomial $G(x)$. Four-bit data 1, 0, 1, and 0, for example, are expressed by a polynomial $D(x) = x^3 + x$. The coding procedure of the data results in

$$D(x) \cdot x^3 = x^6 + x^4 = (x^3 + x + 1)(x^3 + 1) + x + 1, \tag{3.48}$$

$$C(x) = x + 1, \tag{3.49}$$

$$W(x) = D(x) \cdot x^3 + C(x) = x^6 + x^4 + x + 1. \tag{3.50}$$

Thus, three check bits 0, 1, and 1 are added after the data bits. The check bits for each data are calculated by the similar coding procedure, resulting in Table 3.9. Figure 3.4 shows that the code composed of $\boldsymbol{w}_0$–$\boldsymbol{w}_{15}$ is a cyclic code.

The generator matrix of this code is calculated as follows. The check bits for data

$$\boldsymbol{a}_0 = \begin{pmatrix} 1 \\ 0 \\ 0 \\ 0 \end{pmatrix}, \boldsymbol{a}_1 = \begin{pmatrix} 0 \\ 1 \\ 0 \\ 0 \end{pmatrix}, \boldsymbol{a}_2 = \begin{pmatrix} 0 \\ 0 \\ 1 \\ 0 \end{pmatrix}, \text{ and } \boldsymbol{a}_3 = \begin{pmatrix} 0 \\ 0 \\ 0 \\ 1 \end{pmatrix} \text{ are } \boldsymbol{c}_0 = \begin{pmatrix} 1 \\ 0 \\ 1 \end{pmatrix},$$

$$\boldsymbol{c}_1 = \begin{pmatrix} 1 \\ 1 \\ 1 \end{pmatrix} \boldsymbol{c}_2 = \begin{pmatrix} 1 \\ 1 \\ 0 \end{pmatrix}, \text{ and } \boldsymbol{c}_3 = \begin{pmatrix} 0 \\ 1 \\ 1 \end{pmatrix},$$

Table 3.9 Code words of the cyclic code generated by $G(x) = x^3 + x + 1$

Code word	w_0	w_1	w_2	w_3	w_4	w_5	w_6	w_7	w_8	w_9	w_{10}	w_{11}	w_{12}	w_{13}	w_{14}	w_{15}
Data bits	0	0	0	0	0	0	0	0	1	1	1	1	1	1	1	1
	0	0	0	0	1	1	1	1	0	0	0	0	1	1	1	1
	0	0	1	1	0	0	1	1	0	0	1	1	0	0	1	1
	0	1	0	1	0	1	0	1	0	1	0	1	0	1	0	1
Check bits	0	0	1	1	1	1	0	0	1	1	0	0	0	0	1	1
	0	1	1	0	1	0	0	1	0	1	1	0	1	0	0	1
	0	1	0	1	1	0	1	0	1	0	1	0	0	1	0	1

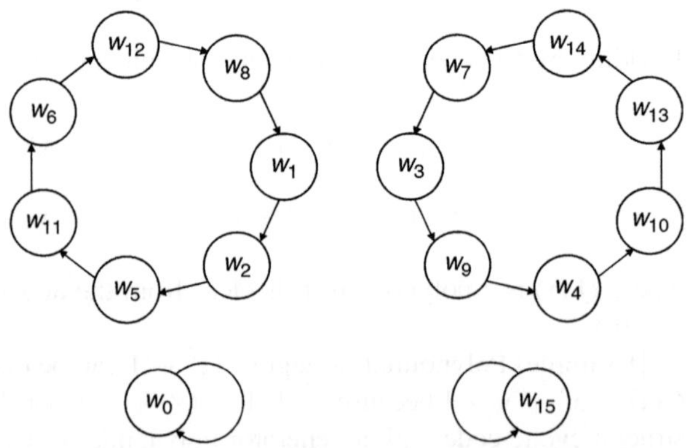

Fig. 3.4 Relationship among code words shown in Table 3.9. The code words w_0–w_{15} are defined in Table 3.9. Each arrow indicates cyclic replacement

respectively. Therefore, the generator matrix is given by

$$G = \begin{pmatrix} 1 & 0 & 0 & 0 \\ 0 & 1 & 0 & 0 \\ 0 & 0 & 1 & 0 \\ 0 & 0 & 0 & 1 \\ 1 & 1 & 1 & 0 \\ 0 & 1 & 1 & 1 \\ 1 & 1 & 0 & 1 \end{pmatrix}. \tag{3.51}$$

The upper part of $\boldsymbol{G}$ is a unit matrix and the lower part is composed of the check bits for the respective column vector of the unit matrix. From (3.51) the check matrix is calculated as

$$\boldsymbol{H} = \begin{pmatrix} 1 & 1 & 1 & 0 & 1 & 0 & 0 \\ 0 & 1 & 1 & 1 & 0 & 1 & 0 \\ 1 & 1 & 0 & 1 & 0 & 0 & 1 \end{pmatrix}. \tag{3.52}$$

The matrix in (3.52) is composed of all the column vectors except for zero vector just like (3.17) though the orders of columns are different. Therefore this cyclic code is a Hamming code, called *cyclic Hamming code*.

3.4.8 Nonbinary Code

The codes described so far are binary codes, where each bit is "0" or "1." However, nonbinary codes are more suitable for multilevel memories. Since the memories store plural bits in a memory cell, all the bits in a cell may be lost at a single incident of soft error. Nonbinary codes can correct the plural bits in a cell even if they are simultaneously lost. When m bits ($q = 2^m$ levels) of data are stored in a memory cell, the m bits are treated as a digit,[4] and an element of Galois field GF(q). For example, in case of two bits/cell, each information stored in each memory cell is assumed as one of the elements of GF(4) (0, 1, α, or α^2) as shown in Fig. 3.5.

The Hamming distance of a nonbinary code is defined as that of a binary code. The Hamming distance $d(\boldsymbol{v}_1, \boldsymbol{v}_2)$ between two vectors $\boldsymbol{v}_1$ and $\boldsymbol{v}_2$ is the number of different digits when compared digit-by-digit (neglecting how different they are). For example, the Hamming distance between

[4]In case of nonbinary code, the term "digit" is used instead of "bit," "data digit" instead of "data bit" and "check digit" instead of "check bit."

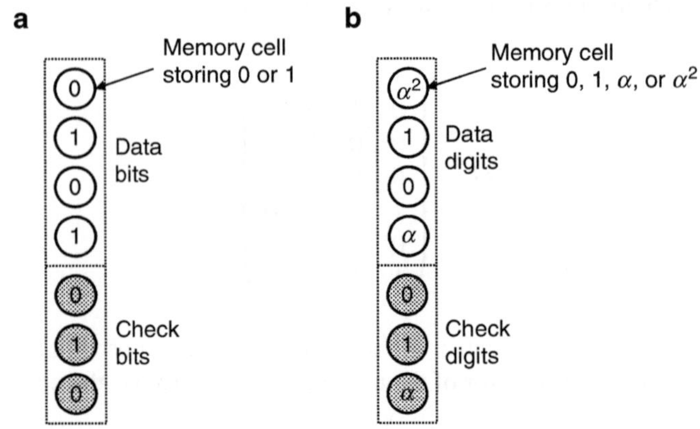

Fig. 3.5 Principle of (**a**) binary code and (**b**) nonbinary (quaternary) code

$$v_1 = \begin{pmatrix} 0 \\ 1 \\ \alpha \\ \alpha^2 \\ 0 \end{pmatrix}$$

and

$$v_2 = \begin{pmatrix} 0 \\ \alpha \\ \alpha \\ \alpha^2 \\ \alpha^2 \end{pmatrix}$$

is $d(\boldsymbol{v}_1, \boldsymbol{v}_2) = 2$. The minimum distance of a nonbinary code is similarly defined as that of binary codes. The minimum distance required for error detection/correction is also the same as the case of binary codes.

Let us calculate the lower bound of required check digits for SEC using Fig. 3.1b. The data-digit count and the code length are assumed as k and n, respectively. The number of single-digit-error cases of a code word $\boldsymbol{w}_1$ is equal to $n(q - 1)$. Because there are n cases to select the digit that errs and there are $(q - 1)$ error patterns of the digit. Therefore, the number of vectors in C_1 (Hamming distance from $\boldsymbol{w}_1$ is within one) is $n(q - 1) + 1$, including no-error case. Since the number of code words is q^k, the total number of vectors in each circle around each code word is $q^k\{n(q - 1) + 1\}$. On the other hand, the number of n-dimensional vectors is q^n. Thus, the following inequality must be satisfied instead of (3.24):

$$q^k\{n(q-1)+1\} \leqslant q^n, \quad q^{n-k} \geqslant n(q-1)+1. \tag{3.53}$$

In the case of DEC, each circle includes (1) 1 vector of no-error case, (2) $n(q-1)$ vectors of single-error cases, and (3) $\binom{n}{2} \cdot (q-1)^2 = n(n-1)(q-1)^2/2$ cases of double-error cases. Therefore, the inequality (3.53) is replaced by

$$q^{n-k} \geqslant 1 + n(q-1) + \frac{n(n-1)(q-1)^2}{2}. \tag{3.54}$$

Generally, for t-digit error correction, the following relationship must be satisfied:

$$q^{n-k} \geqslant \sum_{i=0}^{t} \binom{n}{i} (q-1)^i. \tag{3.55}$$

Table 3.10 shows the required check-digit counts for single-error-correction codes of $q = 2$, 4, 8, and 16. The ratio of required check-digit count decreases with the increase in q for the same k. This is because the data-"bit" count is $k \log_2 q$, which increases with q. The size of the coding/decoding circuit, however, generally increases with q as described in the next section.

A SEC nonbinary code is constructed in a similar way as the binary Hamming code. All the column vectors that are linearly independent from each other are arranged to construct a check matrix. This code is called nonbinary Hamming code. "Linearly independent" means that a vector is not a multiple of another vector. For example,

$$v_1 = \begin{pmatrix} 0 \\ 1 \\ 1 \end{pmatrix}$$

and

$$v_2 = \begin{pmatrix} 0 \\ \alpha \\ \alpha \end{pmatrix}$$

are not linearly independent (linearly dependent) because $v_2 = \alpha v_1$. The number of rows of the check matrix is equal to the check-digit count $n - k$, and the number of column is equal to the code length n as described above. The number of $(n - k)$-

Table 3.10 Lower bound of the number of required check digits for SEC q-ary codes. "$q = 2$" means binary code and is identical to Table 3.7

Data-digit count, k		4	8	16	32	64	128	256
q	2	3	4	5	6	7	8	9
	4	3	3	3	4	4	5	5
	8	2	3	3	3	3	4	4
	16	2	2	3	3	3	3	3

dimensional column vectors except for zero vector is $(q^{n-k} - 1)$. Since every $(q - 1)$ vectors are linearly dependent (vectors $\boldsymbol{v}$, $\alpha\boldsymbol{v}$, $\alpha^2\boldsymbol{v}$, …, $\alpha^{q-2}\boldsymbol{v}$ are linearly dependent), there are $(q^{n-k} - 1)/(q - 1)$ linearly independent vectors. Thus, the following relationship is satisfied, which corresponds to (3.31):

$$\frac{q^{n-k}-1}{q-1} = n. \tag{3.56}$$

[Example] The check matrix of a quaternary ($q = 4$) Hamming code with check-digit count of 3 is given by

$$\boldsymbol{H} = \begin{pmatrix} 1 & 1 & 1 & 1 & 1 & 1 & 0 & 0 & 0 & 1 & 1 & 1 & 1 & 1 & 1 & 1 & 1 & 1 & 1 & 0 & 0 \\ 1 & \alpha & \alpha^2 & 0 & 0 & 0 & 1 & 1 & 1 & 1 & 1 & 1 & \alpha & \alpha^2 & \alpha & \alpha & \alpha^2 & \alpha^2 & 0 & 1 & 0 \\ 0 & 0 & 0 & 1 & \alpha & \alpha^2 & 1 & \alpha & \alpha^2 & 1 & \alpha & \alpha^2 & 1 & 1 & \alpha & \alpha^2 & \alpha & \alpha^2 & 0 & 0 & 1 \end{pmatrix}. \tag{3.57}$$

The code length is $(4^3 - 1)/(4 - 1) = 21$, and the data-digit count is $21 - 3 = 18$. Note that the first nonzero element of each column vector is "1." If a data-digit count of 16 is sufficient, this code is shortened by two digits. Deleting, for example, the 16th and 17th columns, we obtain

$$\boldsymbol{H}' = \begin{pmatrix} 1 & 1 & 1 & 1 & 1 & 1 & 0 & 0 & 0 & 1 & 1 & 1 & 1 & 1 & 1 & 1 & 1 & 0 & 0 \\ 1 & \alpha & \alpha^2 & 0 & 0 & 0 & 1 & 1 & 1 & 1 & 1 & 1 & \alpha & \alpha^2 & \alpha & \alpha^2 & 0 & 1 & 0 \\ 0 & 0 & 0 & 1 & \alpha & \alpha^2 & 1 & \alpha & \alpha^2 & 1 & \alpha & \alpha^2 & 1 & 1 & \alpha & \alpha^2 & 0 & 0 & 1 \end{pmatrix}. \tag{3.58}$$

3.5 Coding and Decoding Circuits

In this section, circuits for the coding (adding check bits) and the decoding (checking and correcting errors) of some of the error correcting codes presented in Sect. 3.4 are described.

3.5.1 Coding and Decoding Circuits for Hamming Code

The coding of a Hamming code is to calculate (3.3). In fact, the upper part of the generator matrix $\boldsymbol{G}$ is a unit matrix and no calculation is needed. The calculation necessary for coding is to multiply the data $\boldsymbol{d}$ by $\boldsymbol{P}$, the lower part of $\boldsymbol{G}$:

$$\boldsymbol{w}' = \begin{pmatrix} w_k \\ w_{k+1} \\ \vdots \\ w_{n-1} \end{pmatrix} = \boldsymbol{Pd}. \tag{3.59}$$

[Example] If the generator matrix is expressed by (3.8), the coding procedure is expressed as

$$\boldsymbol{w}' = \begin{pmatrix} w_4 \\ w_5 \\ w_6 \end{pmatrix} = \begin{pmatrix} 1 & 1 & 0 & 1 \\ 1 & 0 & 1 & 1 \\ 0 & 1 & 1 & 1 \end{pmatrix} \boldsymbol{d}, \tag{3.60}$$

that is,

$$\begin{aligned} &w_4 = d_0 + d_1 + d_3 \mod 2, \quad w_5 = d_0 + d_2 + d_3 \mod 2, \\ &w_6 = d_1 + d_2 + d_3 \mod 2. \end{aligned} \tag{3.61}$$

These calculations are realized by the circuit in Fig. 3.6.

The decoding procedure consists of two steps. The first step is to calculate the syndrome $\boldsymbol{s}$ from the read data $\boldsymbol{r}$ according to (3.13).

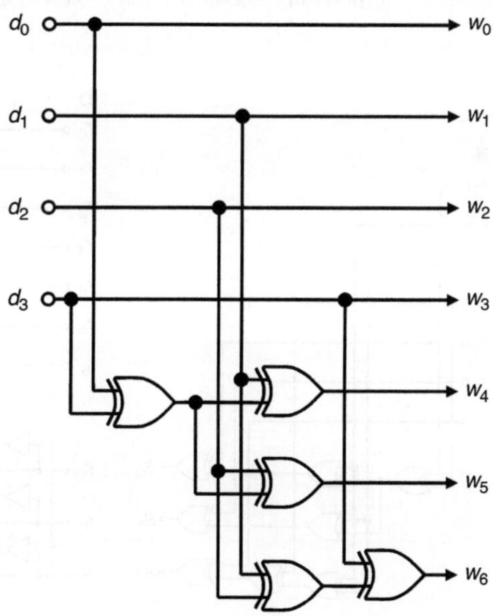

Fig. 3.6 Parallel coding circuit of Hamming code with a code length of 7 and a data-bit count of 4

[Example] If the check matrix is expressed by (3.17), the syndrome is calculated as

$$s = \begin{pmatrix} s_0 \\ s_1 \\ s_2 \end{pmatrix} = \begin{pmatrix} 1 & 1 & 0 & 1 & 1 & 0 & 0 \\ 1 & 0 & 1 & 1 & 0 & 1 & 0 \\ 0 & 1 & 1 & 1 & 0 & 0 & 1 \end{pmatrix} \boldsymbol{r}, \tag{3.62}$$

that is,

$$\begin{aligned} &s_0 = r_0 + r_1 + r_3 + r_4 \mod 2, \quad s_1 = r_0 + r_2 + r_3 + r_5 \mod 2, \\ &s_2 = r_1 + r_2 + r_3 + r_6 \mod 2. \end{aligned} \tag{3.63}$$

These calculations are realized by the circuit in Fig. 3.7a.

The second step is analyzing the syndrome and correcting the read data. From (3.11) and (3.13) we obtain

$$s = \boldsymbol{H}\boldsymbol{r} = \boldsymbol{H}(\boldsymbol{w} + \boldsymbol{e}) = \boldsymbol{H}\boldsymbol{e}. \tag{3.64}$$

If $e_i = 1$ and $e_j = 0$ $(i \neq j)$ (only r_i is erroneous), the syndrome is

$$s = \begin{pmatrix} h_{00} & h_{01} & \cdots & h_{0\,n-1} \\ h_{10} & h_{11} & \cdots & h_{1\,n-1} \\ \vdots & \vdots & \ddots & \vdots \\ h_{n-k-10} & h_{n-k-11} & \cdots & h_{n-k-1\,n-1} \end{pmatrix} \begin{pmatrix} 0 \\ \vdots \\ 1 \\ \vdots \\ 0 \end{pmatrix} = \begin{pmatrix} h_{0i} \\ h_{1i} \\ \vdots \\ h_{n-k-1i} \end{pmatrix}. \tag{3.65}$$

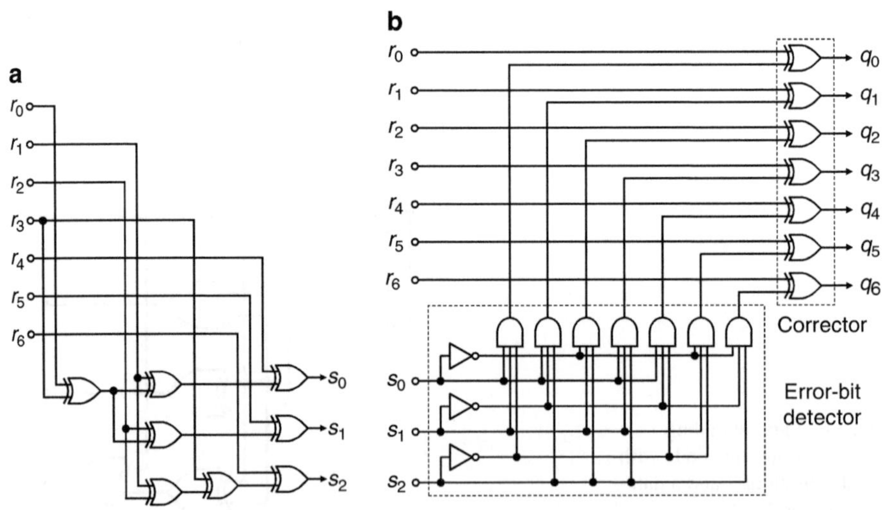

Fig. 3.7 Parallel decoding circuit of Hamming code with a code length of 7 and a data-bit count of 4: (**a**) syndrome generator and (**b**) correction circuit

Thus, the syndrome $\boldsymbol{s}$ is equal to the column vector of $\boldsymbol{H}$ corresponding to the error position. Therefore, $\boldsymbol{s}$ is compared with each column vector of $\boldsymbol{H}$. If $\boldsymbol{s}$ is equal to one of the column vector of $\boldsymbol{H}$, the corresponding read data bit should be corrected.

[Example] Using the above code, if

$$\boldsymbol{s} = \begin{pmatrix} 1 \\ 1 \\ 0 \end{pmatrix}, \begin{pmatrix} 1 \\ 0 \\ 1 \end{pmatrix}, \ldots, \text{ or } \begin{pmatrix} 0 \\ 0 \\ 1 \end{pmatrix},$$

$r_0, r_1, \ldots,$ or r_6 should be corrected, respectively. The circuit to realize this is shown in Fig. 3.7b. In the figure, the error-bit detector judges whether $\boldsymbol{s}$ is equal to each column vector of $\boldsymbol{H}$, and the corrector corrects (inverts) each bit of the read data.

The greater parts of the coding circuit and the syndrome generator are common as shown in Figs. 3.6 and 3.7a. The hardware size can be therefore reduced by using the circuit in Fig. 3.8a or b. The former uses the coding circuit as a part of syndrome generator, while the latter uses the syndrome generator as a coding circuit with some of the input signals fixed to "0".

Let us estimate the circuit size necessary for a Hamming code (code length $= n$, data-bit count $= k$, check-bit count $= m = n - k$). If we use the scheme in Fig. 3.8b, we have only to estimate the decoding circuit. First, the syndrome generator is estimated. As shown in (3.63), each bit of the syndrome is calculated by the "modulo 2" addition of the selected bits among the n-bit read data. The number of selected bits is equal to the number of "1"s in each row of the check matrix. Since there are $(n + 1)/2$ "1"s in each row of the check matrix of a nonshortened Hamming code, $(n - 1)/2$ two-input exclusive OR gates are needed for the "modulo 2" addition. The total number of gates is $m(n - 1)/2$. Next, the correction circuit is composed of m inverters, n m-input AND gates, and n two-input exclusive OR gates, as shown in Fig. 3.7b. Table 3.11 summarizes the number of logic gates for the decoding circuits. These estimations are the upper bounds and the number of gates can be reduced by merging the parts of the gates in the syndrome generator. In Fig. 3.7a, for example, the partial sum $r_0 + r_3$ is commonly used for the two syndrome bits, s_0 and s_1, and the number of exclusive OR gates is 8, which is smaller than 9 shown in Table 3.11. The above estimation is also applied to the case of shortened Hamming code (code length is shortened from n to n'). The number of exclusive OR gates for syndrome generator does not exceed $m(n' - 1)/2$. According to the shortening criteria described in the previous subsection, the ratio of "1"s in the check matrix does not increase by shortening. Figure 3.9 shows the number of MOS transistors when implemented with CMOS circuits. A two-input exclusive OR gate, an inverter, and an m-input AND gate are assumed to consist of 10, 2, and $2m$ transistors, respectively. The exclusive OR gates account for a major part, especially those for the syndrome generator. The number of exclusive OR gates is $m(n - 1)/2$, which is an order of $n \log n$ because $m = \log_2(n + 1)$. In addition to the gate-count reduction described above, it is therefore

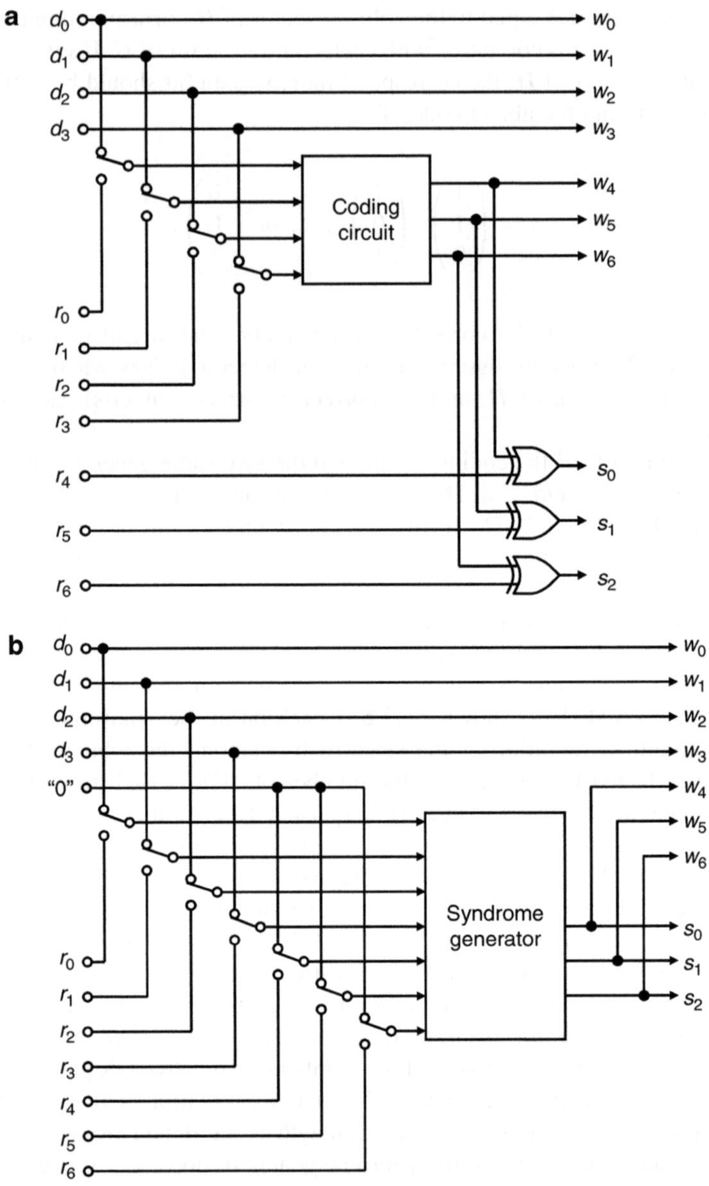

Fig. 3.8 Reducing hardware size of coding/decoding circuits: (**a**) using coding circuit for generating syndrome and (**b**) using syndrome generator for coding

important to realize exclusive-OR function with a small-area circuit, which is discussed in Sect. 3.7. The circuit size increases with the code length *n*, while the ratio of check bits decreases. Therefore, the trade-off between both is needed for ECC design.

Table 3.11 Upper bound of the number of logic gates for the coding/decoding circuits of Hamming codes

Code length, n		7	15	31	63	127	255
Data-bit count, k		4	11	26	57	120	247
Check-bit count, m		3	4	5	6	7	8
Syndrome generator	Two-input exclusive OR	9	28	75	186	441	1,016
Correcting circuit	Two-input exclusive OR	7	15	31	63	127	255
	Inverter	3	4	5	6	7	8
	m-input AND	7	15	31	63	127	255

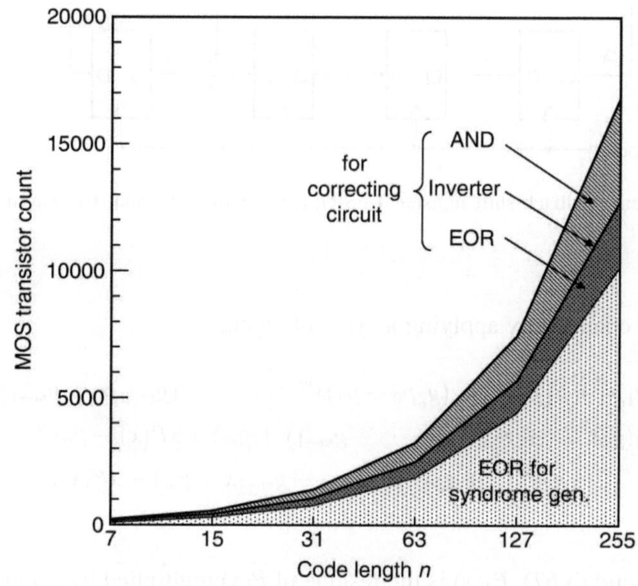

Fig. 3.9 MOS transistor counts for coding/decoding circuits of Hamming codes. An exclusive OR (EOR) gate, an inverter, and an m-input AND gate are assumed to consist of 10, 2, and $2m$ transistors, respectively

3.5.2 Coding and Decoding Circuits for Cyclic Hamming Code

The serial coding and decoding circuits of a cyclic code are realized by a small hardware using the linear-feedback shift register (LFSR) shown in Fig. 3.10a. The circuit is composed of m flip-flops, multipliers, and adders (in fact, exclusive OR gates). The operation of the LFSR is as follows. The data $p_0, p_1, \ldots, p_{m-1}$ stored in the flip-flops are expressed by a polynomial:

$$P(x) = p_0 x^{m-1} + p_1 x^{m-2} + \cdots + p_{m-2} x + p_{m-1}. \tag{3.66}$$

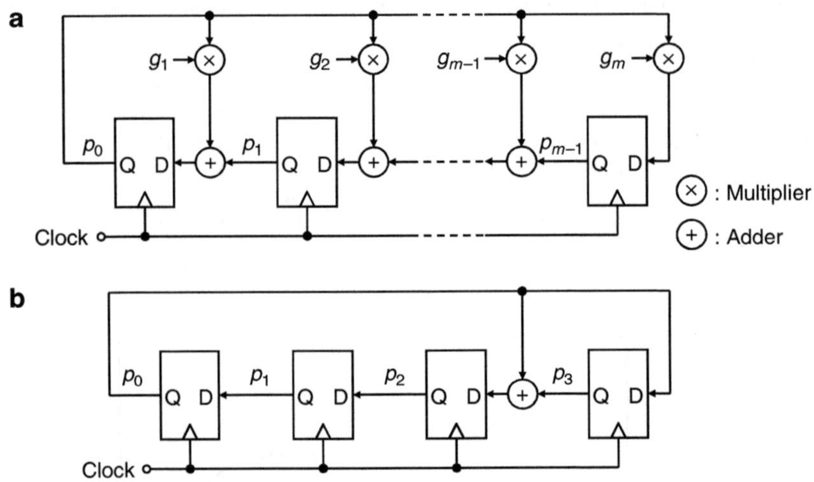

Fig. 3.10 Linear-feedback shift register (LFSR): (**a**) general form and (**b**) case of $G(x) = x^4 + x + 1$

The data are changed by applying a cycle of clock:

$$\begin{aligned}P_1(x) &= (g_1p_0 + p_1)x^{m-1} + (g_2p_0 + p_2)x^{m-2} + \cdots + (g_{m-1}p_0 + p_{m-1})x + g_mp_0\\ &= p_0(g_1x^{m-1} + g_2x^{m-2} + \cdots + g_{m-1}x + g_m) + xP(x) - p_0x^m\\ &= p_0(x^m + g_1x^{m-1} + g_2x^{m-2} + \cdots + g_{m-1}x + g_m) + xP(x).\end{aligned} \tag{3.67}$$

From (3.39) and (3.67), $P_1(x)$ is the residue of $P(x)$ multiplied by x and divided by $G(x)$. In the case of a binary code, each g_i is 0 or 1 and the multipliers are omitted. For example, Fig. 3.10b is the LFSR for $G(x) = x^4 + x + 1$.

The serial coding circuit is shown in Fig. 3.11a. The LFSR is cleared in advance. Data bits $d_0, d_1, \ldots, d_{k-1}$ are inputted in synchronous with the clock signal during the first k cycles and zeros are inputted during the following m cycles. The data stored in the LFSR after i cycles, $P_i(x)$, satisfy the following equations:

$$P_0(x) = 0, \quad P_i(x) = xP_{i-1}(x) + d_{i-1} \mod G(x). \tag{3.68}$$

Since the data stored after n cycles is expressed as

$$P_n(x) = d_0x^{n-1} + d_1x^{n-2} + \cdots + d_{k-1}x^{n-k} \mod G(x), \tag{3.69}$$

the stored bits in the LFSR at this time are the check bits. Another implementation of the coding circuit is shown in Fig. 3.11b. This circuit features that data bits are

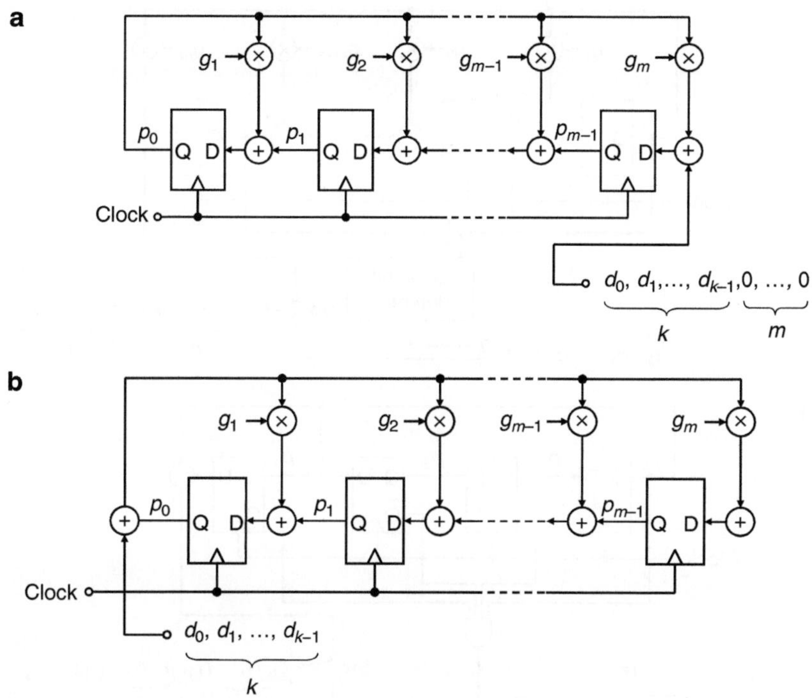

Fig. 3.11 Coding circuits for cyclic code using LFSR: (**a**) inputting data bits from the input of LFSR and (**b**) inputting data bits from the output of LFSR

inputted from the leftmost node, the output terminal of LFSR. The data stored in the LFSR after i cycles, $P_i(x)$, satisfy the following equations:

$$P_0(x) = 0, \quad P_i(x) = xP_{i-1}(x) + d_{i-1}x^m \mod G(x). \tag{3.70}$$

Since the data stored after k cycles is expressed as

$$\begin{aligned} P_k(x) &= d_0x^m \cdot x^{k-1} + d_1x^m \cdot x^{k-2} + \cdots + d_{k-1}x^m \\ &= d_0x^{n-1} + d_1x^{n-2} + \cdots + d_{k-1}x^{n-k} \mod G(x), \end{aligned} \tag{3.71}$$

the stored bits in the LFSR at this time are the check bits. A merit of this circuit is the required cycles are fewer than that of Fig. 3.11a by m cycles. After inputting k data bits during k cycles, the m check bits are serially outputted by shifting the shift register without feedback.

In the case of a shortened cyclic code (code length is shortened from n to n' and data-bit count is from k to k') the first $k - k'$ bits are assumed to be zero. Using the circuit in Fig. 3.11b, we can skip the first $k - k'$ cycles and obtain check bits after k' cycles.

Next, serial decoding circuits using LFSR are discussed. Here, only decoding circuits of cyclic Hamming codes are described. For decoding circuits of other

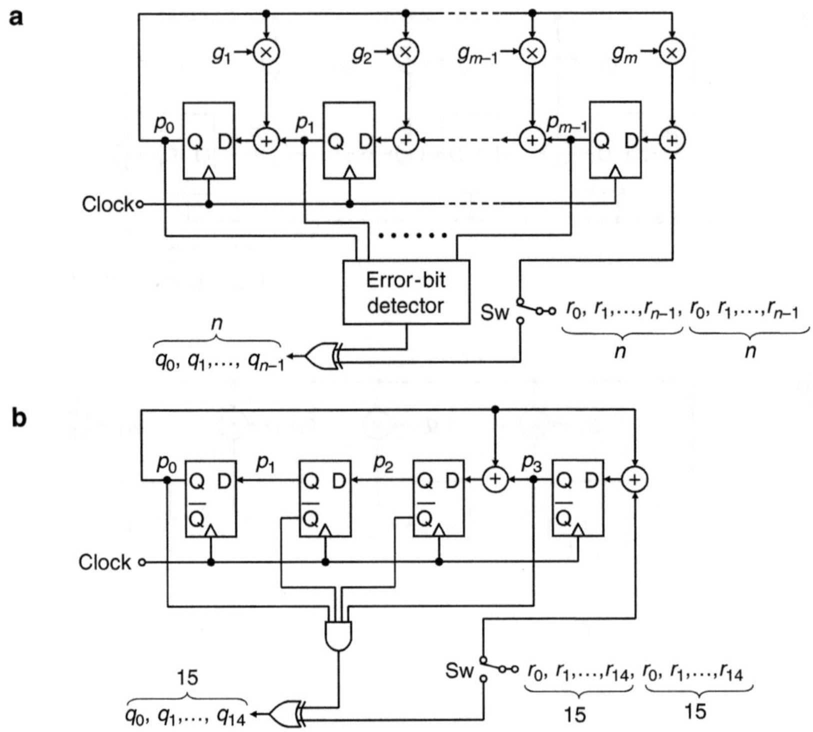

Fig. 3.12 Decoding circuit for cyclic Hamming code using LFSR: (**a**) general form and (**b**) case of $G(x) = x^4 + x + 1$

cyclic codes, refer the textbooks on code theory [1]. The circuit diagram is shown in Fig. 3.12a. The decoding procedure consists of two steps, syndrome generation and data correction, as in the parallel decoding described above. The syndrome is generated by clearing the LFSR in advance, connecting the switch Sw to the upper terminal, and serially inputting $n = k + m$ read data bits. The data stored in the LFSR after i cycles, $P_i(x)$, satisfy the following equations:

$$P_0(x) = 0, \quad P_i(x) = xP_{i-1}(x) + r_{i-1} \mod G(x). \tag{3.72}$$

After n cycles,

$$P_n(x) = r_0x^{n-1} + r_1x^{n-2} + \cdots + r_{n-2}x + r_{n-1} = R(x) \mod G(x) \tag{3.73}$$

is stored in the LFSR. This is the syndrome $S(x)$ because of (3.46). Next, the error position is found using the syndrome. If the first read data bits r_0 is erroneous, the syndrome $s_0, s_1, \ldots, s_{m-1}$ corresponds to x^{n-1} because

$$P_n(x) = S(x) = R(x) = x^{n-1} \mod G(x). \tag{3.74}$$

If bit r_i is erroneous,

$$P_n(x) = S(x) = R(x) = x^{n-1-i} \mod G(x). \tag{3.75}$$

Further shifting the LFSR i cycle results in

$$P_{n+i}(x) = x^i S(x) = x^i R(x) = x^{n-1} \mod G(x). \tag{3.76}$$

Thus, the data stored in LFSR corresponds to x^{n-1}. The correction is therefore performed by connecting the switch Sw to the lower terminal, inputting the read data in synchronous with the clock, and correcting the data bit when the stored data is x^{n-1}. The decoding procedure is completed after $2n$ cycles. If no error, $S(x) = 0$ and the LFSR becomes all zero after n cycles. Since the LFSR remains all zero afterward, no read data bits are corrected.

[Example] Fig. 3.12b shows the decoding circuit for $G(x) = x^4 + x + 1$. Since $x^{n-1} = x^{14} = x^3 + 1 \mod G(x)$, the data bit is corrected when $p_0 = 1$, $p_1 = 0$, $p_2 = 0$, and $p_3 = 1$.

Figure 3.13 shows the case of shortened by three bits. The first three bits r_0, r_1, r_2 are assumed to be "0" and r_3 and the following bits are the read data. The syndrome is generated after fifteen cycles (in fact twelve cycles because the first three cycles are skipped):

$$P_{15}(x) = S(x) \mod G(x). \tag{3.77}$$

If r_3 is erroneous, $S(x) = x^{11}$. The data bit is corrected when the stored data is x^{11}. Since $x^{11} = x^3 + x^2 + x \mod G(x)$, the data bit is corrected when $p_0 = 1$, $p_1 = 1$, $p_2 = 1$, and $p_3 = 0$.

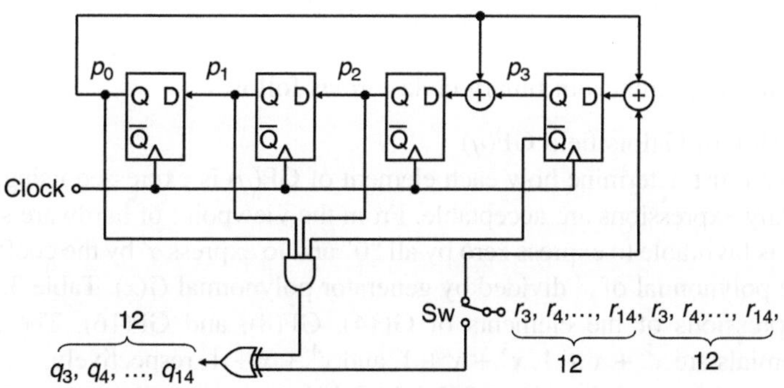

Fig. 3.13 Decoding circuit for shortened cyclic Hamming code with $G(x) = x^4 + x + 1$. Code length is shortened from 15 to 12

Table 3.12 Number of logic gates for the serial decoding circuits of cyclic Hamming codes

Code length, n		7	15	31	63	127	255
Data-bit count, k		4	11	26	57	120	247
Check-bit count, m		3	4	5	6	7	8
LFSR	Flip-flop	3	4	5	6	7	8
	Two-input exclusive OR	2	2	2	2	2	4
Correcting circuit	m-input AND	1	1	1	1	1	1
	Two-input exclusive OR	1	1	1	1	1	1

The polynomials shown in Table 3.3 are used as generator polynomials $G(x)$

The numbers of logic gates necessary for the serial decoding circuits are summarized in Table 3.12. Comparing Figs. 3.11–3.13 with Figs. 3.6–3.8 and Table 3.11 with Table 3.12, it is clear that the circuit size of a serial coding/decoding circuit using LFSR is far smaller than that of a parallel coding/decoding circuit. This is because the circuit size of LFSR is proportional to m (check-bit count), while that of the parallel coding/decoding circuit is approximately proportional to mn (n: code length) as described above. The circuit size of the serial coding/decoding circuit is further reduced by merging the LFSRs of coding and decoding circuits, which is described in Sect. 3.7.

3.5.3 Coding and Decoding Circuits for Nonbinary Code

The coding and decoding procedures of nonbinary codes are described by linear algebra in the same way as those of binary codes. Using the check matrix $\boldsymbol{H}'$ in (3.58), for example, syndrome $\boldsymbol{s}$ is generated from read data $\boldsymbol{r}$ by calculating $\boldsymbol{s} = \boldsymbol{H}'\boldsymbol{r}$, that is,

$$\begin{aligned}
s_0 &= r_0+r_1+r_2+r_3+r_4+r_5+r_9+r_{10}+r_{11}+r_{12}+r_{13}+r_{14}+r_{15}+r_{16},\\
s_1 &= r_0+\alpha r_1+\alpha^2 r_2+r_6+r_7+r_8+r_9+r_{10}+r_{11}+\alpha r_{12}+\alpha^2 r_{13}+\alpha r_{14}+\alpha^2 r_{15}+r_{17},\\
s_2 &= r_3+\alpha r_4+\alpha^2 r_5+r_6+\alpha r_7+\alpha^2 r_8+r_9+\alpha r_{10}+\alpha^2 r_{11}+r_{12}+r_{13}+\alpha r_{14}+\alpha^2 r_{15}+r_{18}.
\end{aligned} \tag{3.78}$$

The different points from binary codes are as follows.

1. Expression of Galois field GF(q)
   First, we must determine how each element of GF(q) is expressed using "0"s and "1"s. Any expressions are acceptable. From the viewpoint of hardware size, however, it is favorable to express zero by all "0" and to express α^i by the coefficients of residue polynomial of x^i divided by generator polynomial $G(x)$. Table 3.13 shows the expressions of the elements of GF(4), GF(8), and GF(16). The generator polynomials are $x^2 + x + 1$, $x^3 + x + 1$, and $x^4 + x + 1$, respectively.
2. Operations between elements of Galois field
   The circuits for addition and multiplication between the elements of Galois field are needed as understood by (3.78). If the expression above is used, an adder is

Table 3.13 Assignment of Galois Field elements: GF(4) ($G(x) = x^2 + x + 1$), GF(8) ($G(x) = x^3 + x + 1$), and GF(16) ($G(x) = x^4 + x + 1$)

GF(4) ($G(x) = x^2 + x + 1$)

	$x1$	$x0$
0	0	0
1	0	1
α	1	0
α^2	1	1

GF(8) ($G(x) = x^3 + x + 1$)

	$x2$	$x1$	$x0$
0	0	0	0
1	0	0	1
β	0	1	0
β^2	1	0	0
β^3	0	1	1
β^4	1	1	0
β^5	1	1	1
β^6	1	0	1

GF(16) ($G(x) = x^4 + x + 1$)

	$x3$	$x2$	$x1$	$x0$
0	0	0	0	0
1	0	0	0	1
γ	0	0	1	0
γ^2	0	1	0	0
γ^3	1	0	0	0
γ^4	0	0	1	1
γ^5	0	1	1	0
γ^6	1	1	0	0
γ^7	1	0	1	1
γ^8	0	1	0	1
γ^9	1	0	1	0
γ^{10}	0	1	1	1
γ^{11}	1	1	1	0
γ^{12}	1	1	1	1
γ^{13}	1	1	0	1
γ^{14}	1	0	0	1

realized by bit-by-bit exclusive OR as shown in Fig. 3.14a or 3.15a. As for a multiplier, a general multiplier circuit is not needed. Multiply-by-constant circuits are sufficient to calculate (3.78). Multiply-by-α and multiply-by-α^2 circuits on GF(4) are shown in Fig. 3.14b, c, respectively. Multiply-by-constant circuits on GF(8) are shown in Fig. 3.15b–g. Confirm that the multiplications of Tables 3.4 and 3.5 are realized by these circuits. An adder and multiply-by-constant circuits on GF(16) are constructed in a similar way, though they are more complex.

3. Error correction

In the case of binary codes, when we find an erroneous bit, we have only to invert the bit for correction. In the case of nonbinary codes, however, we have to know

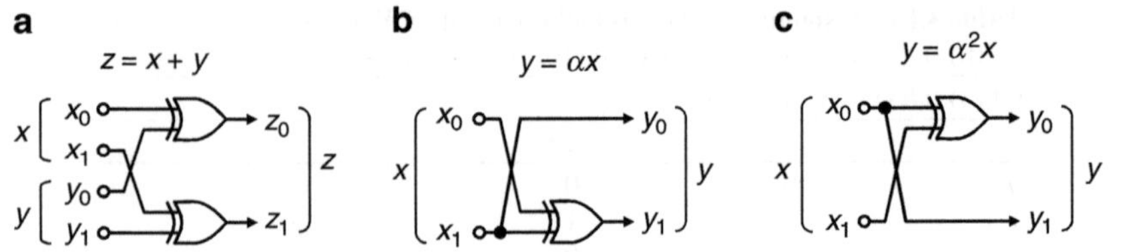

Fig. 3.14 Adder and multipliers on GF(4) ($G(x) = x^2 + x + 1$): (**a**) adder, (**b**) multiply-by-α circuit, and (**c**) multiply-by-α^2 circuit

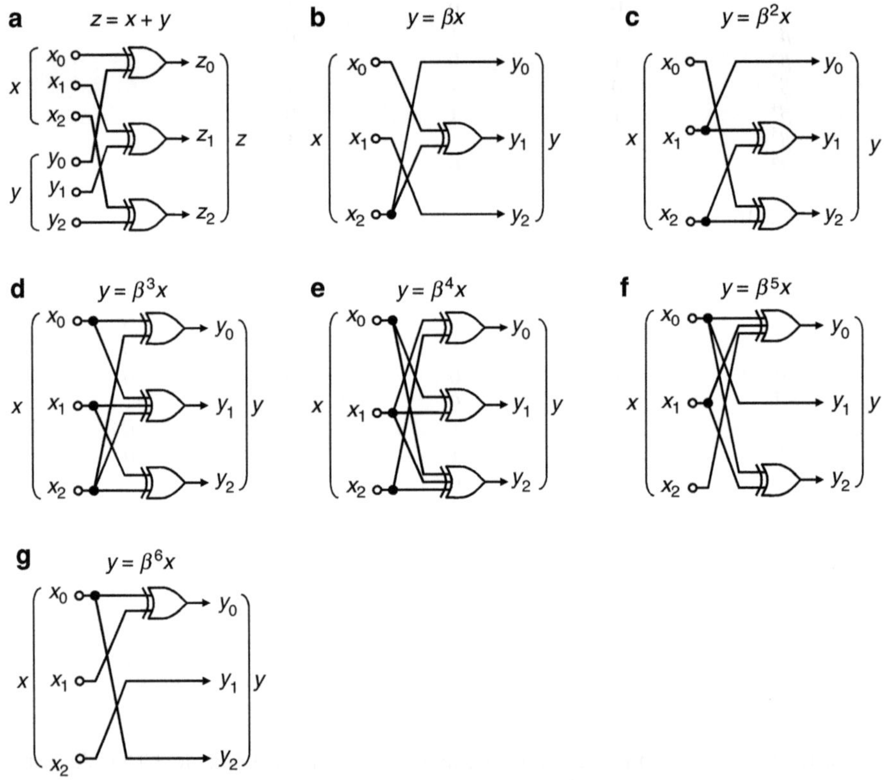

Fig. 3.15 Adder and multipliers on GF(8) ($G(x) = x^3 + x + 1$): (**a**) adder, (**b**) multiply-by-β circuit, (**c**) multiply-by-β^2 circuit, (**d**) multiply-by-β^3 circuit, (**e**) multiply-by-β^4 circuit, (**f**) multiply-by-β^5 circuit, and (**g**) multiply-by-β^6 circuit

not only the error position (which digit is erroneous) but also error size (how the digit is erroneous).

[Example] Let us consider a quaternary code with the check matrix $\boldsymbol{H}'$ of (3.58). The first column of $\boldsymbol{H}'$ is

$$\begin{pmatrix} 1 \\ 1 \\ 0 \end{pmatrix}.$$

If the syndrome is

$$s = \begin{pmatrix} 1 \\ 1 \\ 0 \end{pmatrix},$$

the first read digit r_0 is judged to have an error of size "1" and is added by "1." If

$$s = \begin{pmatrix} \alpha \\ \alpha \\ 0 \end{pmatrix},$$

r_0 is judged to have an error of size "α" and is added by "α." If

$$s = \begin{pmatrix} \alpha^2 \\ \alpha^2 \\ 0 \end{pmatrix},$$

r_0 is judged to have an error of size "α^2" and is added by "α^2". In this way, we must check whether the syndrome $\boldsymbol{s}$ is not only equal to a column vector of $\boldsymbol{H}$, but also equal to a multiple of a column vector. Another correction method is as follows. If $s_0 = s_1 \neq 0$ and $s_2 = 0$, s_0 is added to r_0, if $\alpha s_0 = s_1 \neq 0$ and $s_2 = 0$, s_0 is added to r_1, if $\alpha^2 s_0 = s_1 \neq 0$ and $s_2 = 0$, s_0 is added to r_2, if $s_0 = s_2 \neq 0$ and $s_1 = 0$, s_0 is added to r_3, and so on.

3.6 Theoretical Reduction in Soft-Error and Hard-Error Rates

On-chip ECC is effective for reducing both soft errors and hard errors of memories. The reduction of the error rates by using ECC is mainly determined by the error-correcting capability (SEC, or DEC, etc.), code length, and memory capacity as discussed below.

3.6.1 Reduction in Soft-Error Rate

On-chip ECC is the key for reducing soft errors of random-access memories (RAMs) in the future, especially SRAMs with inherent small signal charge. The soft-error rate (SER) reduction of a RAM by using ECC is calculated as follows [3].

3.6.1.1 Single-Error Correction

Let us consider an M-bit memory. If we use a single-error correcting code with a code length of n and data-bit count of k, the number of code words included in the memory is $W = M/k$. Since soft errors are rare and random phenomena, the probability of soft-error occurrence in an ECC code word during period T is assumed to follow Poisson's distribution:

$$P(i) = \frac{(n\varepsilon T)^i \exp(-n\varepsilon T)}{i!}, \tag{3.79}$$

where $P(i)$ is the probability of i soft errors occur during period T, n is the code length, and ε is the SER of a single memory cell. Therefore, the probability of correctable error (i.e., one error maximum) is

$$P(0) + P(1) = (1 + n\varepsilon T)\exp(-n\varepsilon T). \tag{3.80}$$

The probability of correctable errors of the entire memory chip is

$$\{P(0) + P(1)\}^W = \exp(-ET), \tag{3.81}$$

where E is the SER of the entire chip. From (3.80) and (3.81), we obtain

$$E = n\varepsilon W - \frac{W}{T}\ln(1 + n\varepsilon T) \cong \frac{Wn^2\varepsilon^2 T}{2}, \tag{3.82}$$

using the Taylor expansion $\ln(1 + x) \approx x - x^2/2$ for small x. On the other hand, the soft error rate of the chip without error correction E_0 is given by

$$E_0 = Wk\varepsilon, \tag{3.83}$$

where k is the number of data bits in an ECC code word. From (3.82) and (3.83) we obtain

$$E \cong \frac{n^2 T}{2k^2 W} \cdot E_0^2. \tag{3.84}$$

Thus, E is proportional to the square of E_0. Figure 3.16 shows the SER reduction using a code of $n = 136$ and $k = 128$ [4]. Even if SER without ECC is as high as 10^6 FIT (one upset per 1,000 h) and the errors are not corrected at all during 10 years ($T = 10$ years), the SER is improved by about four orders of magnitude through ECC. In the case of DRAMs, error correction is usually performed every refresh cycle to prevent the accumulation of errors [5, 6]. If periodic error correction (one ECC code word every 7.8 μs, $T = 7.8$ μs × W), the resulting SER

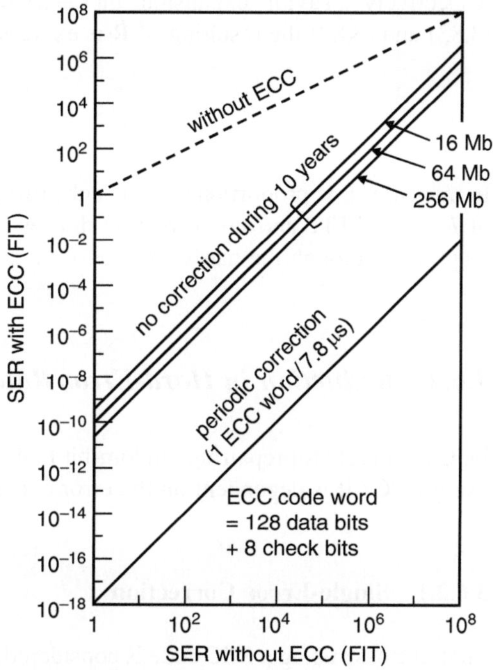

Fig. 3.16 Soft-error rate (SER) reduction through on-chip ECC. Reproduced from [4] with permission; © 2010 IEEE

becomes as low as 10^{-6} FIT. Thus, periodic correction is effective for reducing SER of not only DRAMs, but also SRAMs. Measurement results of SER reduction is presented in Sect. 3.7.

3.6.1.2 Double-Error Correction

If a DEC code is used, the probability of correctable error (i.e., two errors maximum) in an ECC code word is expressed as

$$P(0) + P(1) + P(2) = \left\{1 + n\varepsilon T + \frac{(n\varepsilon T)^2}{2}\right\} \exp(-n\varepsilon T). \tag{3.85}$$

Therefore, (3.81) and (3.82) are replaced by

$$\{P(0) + P(1) + P(2)\}^W = \exp(-ET), \tag{3.86}$$

and

$$E = n\varepsilon W - \frac{W}{T} \ln\left\{1 + n\varepsilon T + \frac{(n\varepsilon T)^2}{2}\right\} \cong \frac{Wn^2\varepsilon^2 T}{6}, \tag{3.87}$$

respectively (Taylor expansion $\ln(1 + x) \approx x - x^2/2 + x^3/3$ was used). From (3.83) and (3.87) the resulting SER is expressed as:

$$E \cong \frac{n^3T^2}{6k^3W^2} \cdot E_0^3. \tag{3.88}$$

In this case, E is proportional to the cube of E_0. The SER reduction rate in the case of $E_0 = 10^6$ FIT and no correction during 10 years is as large as 7–9 orders of magnitude, though the hardware size of the circuit is several times that of SEC.

3.6.2 *Reduction in Hard-Error Rate*

ECC is suitable for repairing random-bit faults. How many faults can be repaired by using ECC? It is dependent on the error correcting capability of the code used.

3.6.2.1 Single-Error Correction

First, the case using a SEC code is considered. To repair all the faulty bits in a chip, any two faulty bits must be located in different code words. The probability $Y(K)$ that a memory chip with K faulty bits is repairable is calculated as follows.

Let us consider the memory contains W code words and each code word consists of n bits (i.e., the code length is n), including check bits. The first faulty bit may be located in any code word. The second faulty bit must be located in a code word that is different from the code word of the first faulty bit. The probability is $n(W - 1)/(nW - 1)$ because there are $n(W - 1)$ bits in the remaining $(W - 1)$ code words, while there are $(nW - 1)$ bits except for the first faulty bits. The third faulty bit must be located in a code word that is different from the first or the second code word. The probability is $n(W - 2)/(nW - 2)$. Continuing similar calculations to the Kth faulty bit and multiplying all the probabilities result in

$$\begin{aligned} Y(K) &= \frac{n(W-1)}{nW-1} \cdot \frac{n(W-2)}{nW-2} \cdot \cdots \cdot \frac{n(W-K+1)}{nW-K+1} \\ &= \frac{n^K(W-1)!(nW-K)!}{(nW-1)!(W-K)!}. \end{aligned} \tag{3.89}$$

Figure 3.17 shows the repairable probability $Y(K)$ for 16-Mbit to 1-Gbit memories using a single-error correcting code of $n = 136$ (128 data bits and 8 check bits). The number of repairable faulty bits is far smaller than the potential possibility. In the case of 256-Mbit memory, for example, more than two million faults may be potentially repaired because $W = 2{,}097{,}152$. However, far fewer (around 2,000) faults can be actually repaired. This is because the probability of "fault collision"

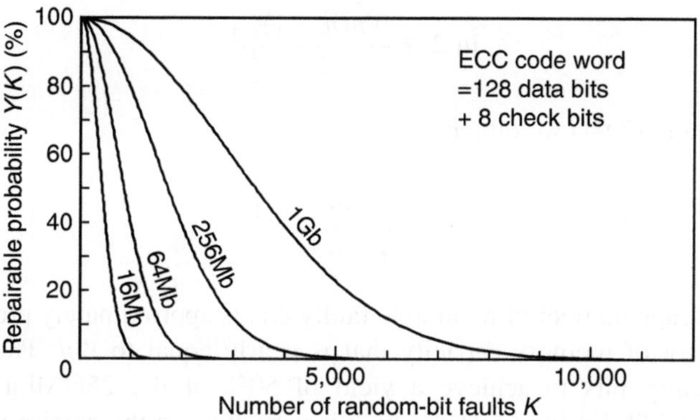

Fig. 3.17 Repairable probability of random-bit faults through on-chip ECC using a single-error correcting code

(two or more faults being located in a code word) is unexpectedly high.[5] Let us derive a simpler approximation to estimate the number of repairable bits. If $K \ll Wn$, the fault distribution is assumed to be a binomial distribution. That is, the probability that a code word has i faulty bits is expressed as

$$P(i) = \binom{n}{i} p^i (1-p)^{n-i}, \tag{3.90}$$

$$p = \frac{K}{Wn}, \tag{3.91}$$

where p is bit-error rate. Since the probability of a code word being repairable (having at most one faulty bit) is $P(0) + P(1)$, the yield is given by

$$Y = \{P(0) + P(1)\}^W = [(1-p)^{n-1}\{1 + (n-1)p\}]^W. \tag{3.92}$$

The number of repairable faulty bits to achieve a yield of 50% is estimated by substituting $Y = 0.5$ into (3.92):

$$-\ln 2 = W[(n-1)\ln(1-p) + \ln\{1 + (n-1)p\}]. \tag{3.93}$$

Using Taylor expansion $\ln(1 + x) \approx x - x^2/2$ and neglecting the terms of p^3 or higher result in

[5] This is known as "birthday problem (birthday paradox)." When K persons are randomly selected, the probability of at least two of them having the same birthday is unexpectedly high. It exceeds 50% for $K \geq 23$.

$$\ln 2 \cong \frac{Wn(n-1)}{2}p^2. \tag{3.94}$$

From (3.91) and (3.94) we obtain

$$K \cong \sqrt{\frac{2Wn \ln 2}{n-1}}. \tag{3.95}$$

Thus, the average number of repairable faulty bits is approximately proportional to the square root of memory capacity that is nearly equal to Wn. The number of repairable faulty bits to achieve a yield of 50% of the 256-Mbit memory is estimated as 1,711.4 using (3.94),[6] which is very close to the precise value, 1,711.

3.6.2.2 Double-Error Correction

In the case of using a double-error correcting code, it is possible to derive an analytical solution, but it is too complex. So let us utilize the approximation using binomial distribution again. The probability of a code word being repairable (having at most two faulty bits) is $P(0) + P(1) + P(2)$ in this case. Therefore, (3.92)–(3.94) are replaced by

$$\begin{aligned} Y &= \{P(0) + P(1) + P(2)\}^W \\ &= \left[(1-p)^{n-2}\left\{1 + (n-2)p + \frac{(n-1)(n-2)}{2}p^2\right\}\right]^W, \end{aligned} \tag{3.96}$$

$$\begin{aligned} -\ln 2 = W\Bigg[&(n-2)\ln(1-p) \\ &+ \ln\left\{1 + (n-2)p + \frac{(n-1)(n-2)}{2}p^2\right\}\Bigg], \end{aligned} \tag{3.97}$$

$$\ln 2 \cong \frac{Wn(n-1)(n-2)}{6}p^3, \tag{3.98}$$

respectively (Taylor expansion $\ln(1 + x) \approx x - x^2/2 + x^3/3$ was used and the terms of p^4 or higher were neglected). Figure 3.18 shows the repairable probability $Y(K)$ for 16-Mbit to 1-Gbit memories using a double-error correcting code of $n = 144$ (128 data bits and 16 check bits). The number of repairable faulty bits is more than 10 times that using a SEC code. From (3.91) and (3.98) we obtain

[6] In the case of birthday problem, substituting $W = 365$ and $n = \infty$ into (3.95) results in $K = 22.5$.

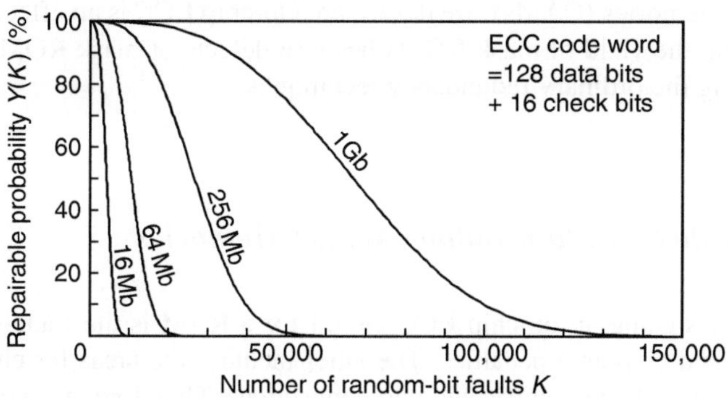

Fig. 3.18 Repairable probability of random-bit faults through on-chip ECC using a double-error correcting code

$$K \cong \sqrt[3]{\frac{6W^2n^2 \ln 2}{(n-1)(n-2)}}. \tag{3.99}$$

In this case, the number of repairable faulty bits is approximately proportional to the memory capacity to the power 2/3.

3.7 Application of ECC

This section discloses some examples of on-chip ECC techniques applied to various kinds of memory LSIs and discusses various practical problems at the design of ECC circuits. Although on-chip ECC techniques are very effective for both soft errors and hard errors of memory LSIs as described in the previous section, they produce significant chip-area and access-time penalties. Each kind of memory LSI has its own optimum solution to minimize the penalties. First, application to random-access memories, including DRAMs and SRAMs as well as "random-access" nonvolatile memories, is described. The percentage of check bits decreases as the code length increases, while the size of coding/decoding circuit and its delay time increase. Therefore, the key to designing the ECC circuits for these memories is the trade-off between access-time and chip-area penalties. Problems inherent to random-access memories, partial-write problem, and startup problem are also discussed. Second, application to serial-access memories, such as "serial-access" nonvolatile memories, is described. Cyclic codes are just the codes for these kinds of memories. The size of coding/decoding circuit is dramatically reduced by using the cyclic code. Third, application to multilevel-storage memories is described. Although multilevel storage is effective for realizing large-capacity memories, the memories are more susceptible to errors than binary memories due to small difference between storage levels. In addition, a single incident of error may destroy all the data bits stored in a cell. The solutions to these problems are explained in this section. Finally, application to other memories, mask ROMs, and content addressable memories (CAMs), are described. On-chip ECC is an effective scheme for improving the yield of mask ROMs because defects of mask ROMs cannot be repaired using the ordinary redundancy techniques.

3.7.1 Application to Random-Access Memories

The key to designing an on-chip ECC circuit for a RAM is the trade-off between access-time and chip-area penalties. The latter includes the areas for check bits for ECC, coding/decoding circuitry, and interconnections. Therefore, it is not always the best solution to minimize the ratio of check bits. This is because an error-correcting code with smaller check-bits ratio generally requires more data bits (a longer code word) and a larger coding/decoding circuit as described in Sect. 3.4. The error correcting codes proposed for RAMs so far are broadly divided into two categories, (extended) Hamming codes and bidirectional parity codes. The former can minimize the check-bit ratio, while the latter can realize a smaller coding/decoding circuit.

3.7.1.1 ECC Using Bidirectional Parity Code

The schematic of the bidirectional parity code is shown in Fig. 3.2 [5, 6]. A code word consists of $k_x k_y$ data bits, k_x horizontal parity (H-parity) bits, k_y vertical parity (V-parity) bits, and one horizontal–vertical parity (HV-parity) bit as described in Sect. 3.4. The code length is therefore $(k_x + 1)(k_y + 1)$. The coding (generation of parity bits) procedure is as follows. First, the data bits are logically (not physically) arranged as a matrix of k_x times k_y as shown in Fig 3.2. Each H-parity bit h_i is determined so that each row contains even "1"s and each V-parity bit v_j is determined so that each column contains even "1"s:

$$d_{i0} + d_{i1} + \cdots + d_{i\,k_y-1} + h_i = 0 \mod 2, \tag{3.100}$$

$$d_{0j} + d_{1j} + \cdots + d_{k_x-1\,j} + v_j = 0 \mod 2, \tag{3.101}$$

where d_{ij} is the ith row, jth column element of the matrix. Finally, the HV-parity bit z is determined so that

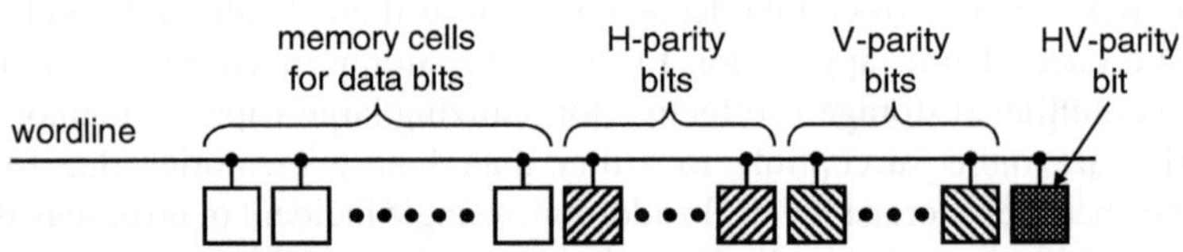

Fig. 3.19 Physical cell arrangement for bidirectional parity code shown in Fig. 3.2 [6]

$$h_0 + h_1 + \cdots + h_{k_x-1} + z = 0 \mod 2, \tag{3.102}$$

$$v_0 + v_1 + \cdots + v_{k_y-1} + z = 0 \mod 2. \tag{3.103}$$

The $(k_x + 1)(k_y + 1)$ bits forming a code word are arranged on a wordline as shown in Fig. 3.19. ECC procedure is as follows. When data d_{ij} is read out, parity checks are performed within the horizontal and vertical groups, H_i and V_j, that d_{ij} belongs to:

$$P_{\mathrm{H}} = d_{i0} + d_{i1} + \cdots + d_{i\,k_y-1} + h_i \mod 2, \tag{3.104}$$

$$P_{\mathrm{V}} = d_{0j} + d_{1j} + \cdots + d_{k_x-1\,j} + v_j \mod 2. \tag{3.105}$$

In case of no error, both P_{H} and P_{V} should be 0. If $P_{\mathrm{H}} = P_{\mathrm{V}} = 1$, d_{ij} is assumed to be erroneous and is corrected. However, if $P_{\mathrm{H}} = 1$ and $P_{\mathrm{V}} = 0$, another bit in the same H_i group is assumed to be erroneous. If $P_{\mathrm{H}} = 0$ and $P_{\mathrm{V}} = 1$, another bit in the same V_j group is assumed to be erroneous. In these cases, d_{ij} is not corrected. Error checking and correcting of the parity bits are performed with the same procedure. H-parity bit h_i, for example, is corrected if both

$$P_{\mathrm{H}} = d_{i0} + d_{i1} + \cdots + d_{i\,k_y-1} + h_i \mod 2 \tag{3.106}$$

and

$$P_{\mathrm{V}} = h_0 + h_1 + \cdots + h_{k_x-1} + z \mod 2 \tag{3.107}$$

are equal to 1. Note that only $(k_x + k_y - 1)$ bits among $(k_x + 1)(k_y + 1)$ bits of a code word are necessary for the procedure. This is one of the features of the bidirectional parity code, while Hamming code described below requires all the bits of a code word for error correction. The HV-parity bit z may be omitted [5]. In this case, however, parity bits cannot be corrected.

The on-chip ECC circuits using the bidirectional parity code applied to 256 kb to 1 Mb DRAMs are shown in Fig. 3.20 [5, 6]. When a wordline is selected, all the bits in a code word are simultaneously read out from the memory cells to the datalines. The target bit is selected by the multiplexer. On the other hand, the bits necessary for the correction of the target bit are selected by the selectors and are sent to the H- and V-parity checkers. Selecting signals $H, \bar{H}, V$ and $\bar{V}$ specify the type of the target bit. The combination of $\bar{H}$ and $\bar{V}$ specifies correcting a data bit, H and $\bar{V}$ an H-parity bit, $\bar{H}$ and V a V-parity bit, and H and V the HV-parity bit. The target bit is selected by the multiplexer. If both the outputs of the parity check circuits are "1" as mentioned above, the target bit is inverted before outputting or rewriting.

Periodical error correction of data and parity bits is effective for avoiding bit-error accumulation as described in Sect. 3.6. In case of DRAMs, the error correction is usually performed every refresh operation [6]. The on-chip address counter

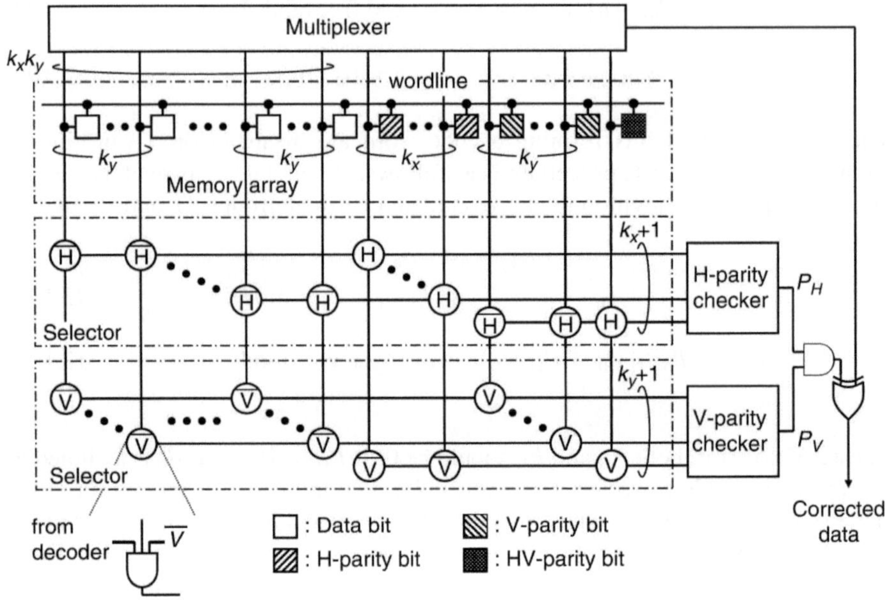

Fig. 3.20 Logic diagram of error correction circuit using bidirectional parity code. Reproduced from [6] with permission; © 2010 IEEE

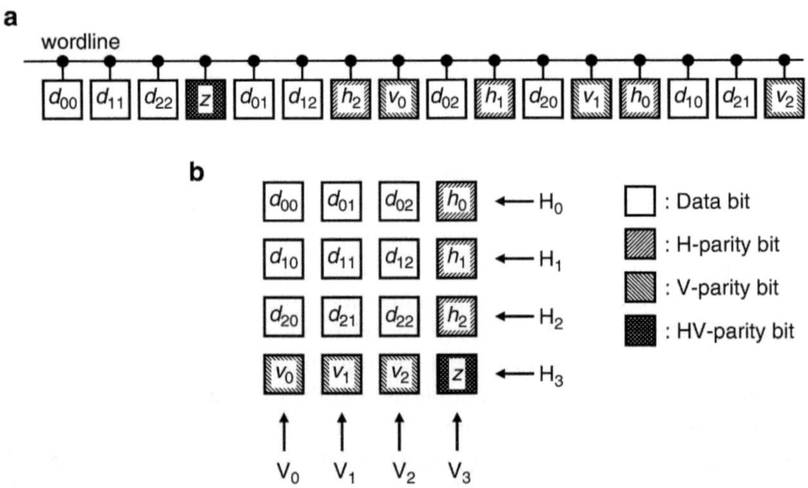

Fig. 3.21 Improved cell assignment for bidirectional parity code: (**a**) physical arrangement and (**b**) logical arrangements. Reproduced from [7] with permission; © 2010 IEEE

generates the selecting signals (*H* and *V*) and column address signals in addition to row address signals during refresh cycles. Thus, all data bits and parity bits are successively scanned.

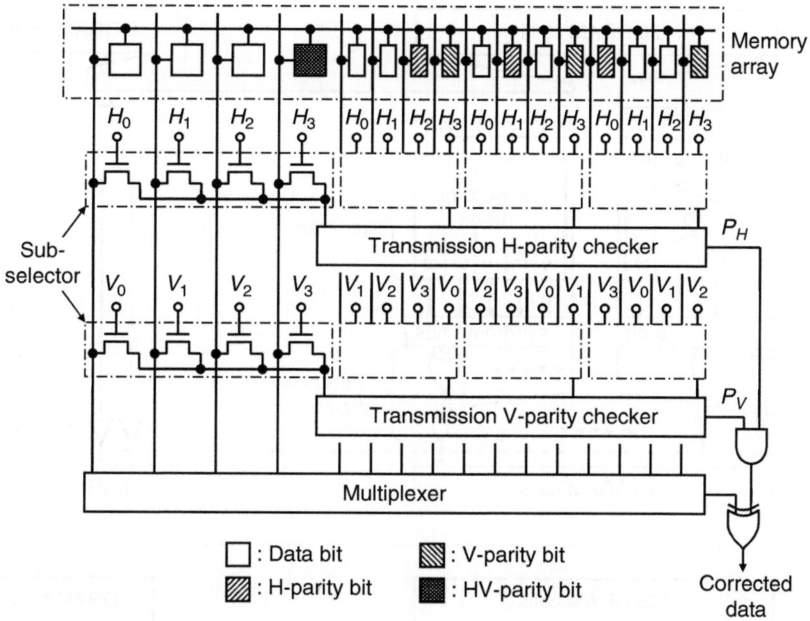

Fig. 3.22 Logic diagram of error correction circuit using bidirectional parity code and improved cell assignment. Reproduced from [7] with permission; © 2010 IEEE

Figure 3.21 shows an improved cell arrangement for reducing the access-time and area penalties due to the interconnections between the selectors and the parity checkers [7]. The case of $m = n = 3$ is shown for simplicity. This arrangement features that the data and parity bits are diagonally arranged in the matrix. Therefore, each cell in a group of adjacent four cells (0–3, 4–7, 8–11, or 12–15) belongs to a different row and a different column. In other word, the matrix in Fig. 3.21b is a Latin square. The on-chip ECC circuits using the cell assignment are shown in Fig. 3.22. Both the H and V selectors are, respectively, divided into four subselectors. The selecting signals H_0–H_3 for the H-selector are supplied repeatedly every four bits, while V_0–V_3 for the V-selector are input with one bit shifted every four bits. The parity checker is arranged close to the respective selector. Therefore, this architecture minimizes the length of the interconnections between the selector and the parity checker. It is reported that an access-time penalty of below 5 ns and a chip-area penalty of 10% have been achieved using this architecture of $k_x = k_y = 16$ [7].

3.7.1.2 ECC Using (Extended) Hamming Code

Hamming code and extended Hamming code realize a small check-bit ratio as described in Sect. 3.4. The coding/decoding circuits for them, however, are larger

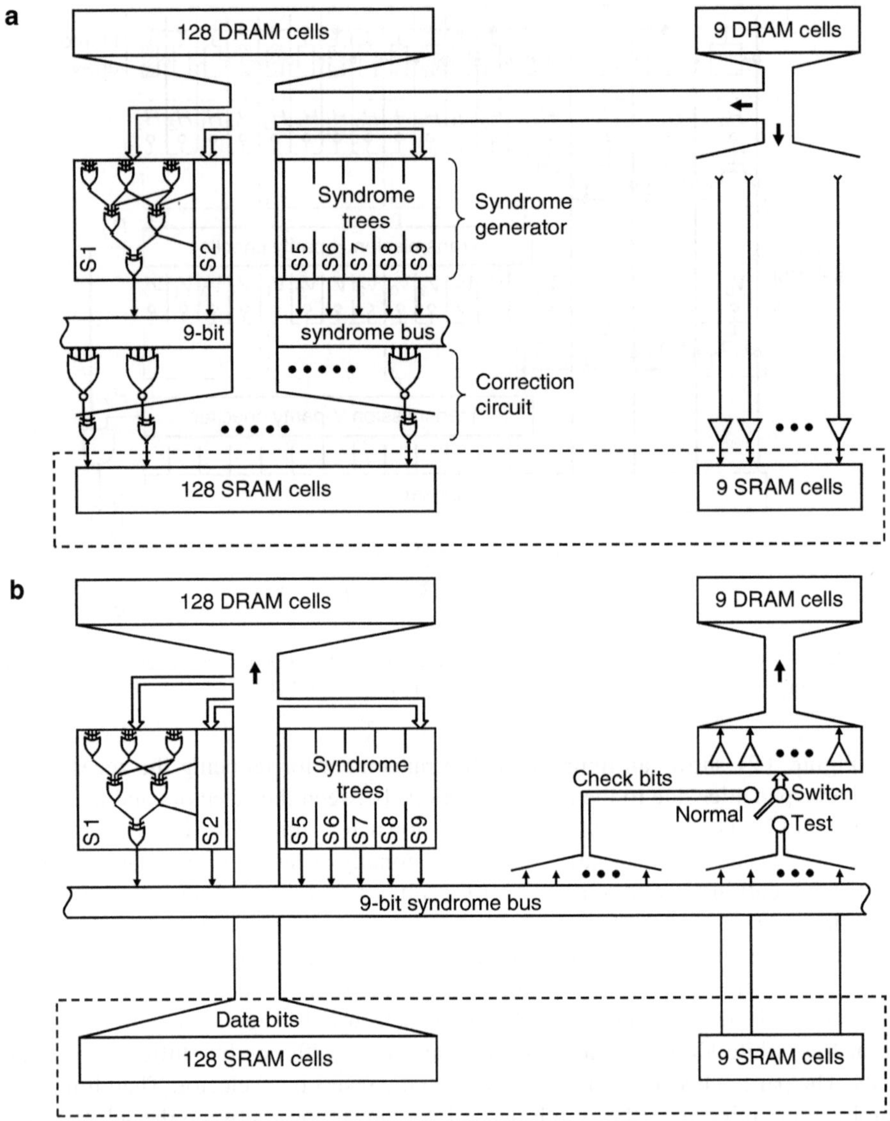

Fig. 3.23 ECC circuit for a 16-Mbit DRAM: (**a**) read operation and (**b**) write operation. Reproduced from [8] with permission; © 2010 IEEE

than that for the bidirectional parity code. The reduction of circuit size and delay time is therefore the key to designing the on-chip ECC circuit using them.

The schematic of an on-chip ECC circuit for a 16-Mbit DRAM using is shown in Fig. 3.23 [8]. A Hsiao code [2] of a data-bit count of 128 and a check-bit count of 9 is used for SEC–DED. The read and write operations are shown in Fig. 3.23a, b,

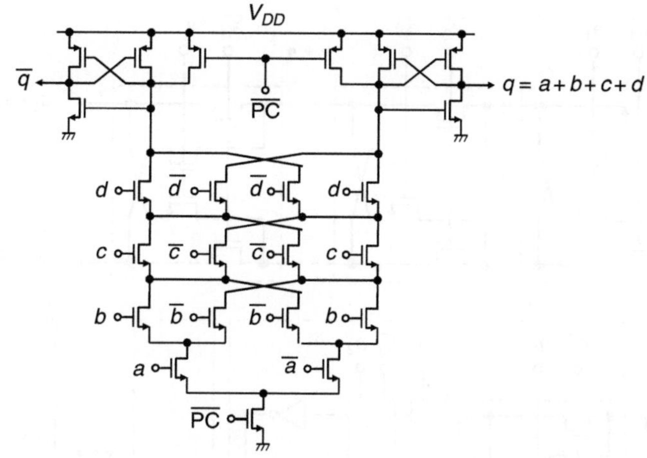

Fig. 3.24 Clocked DCVS four-input exclusive-OR circuit [9]

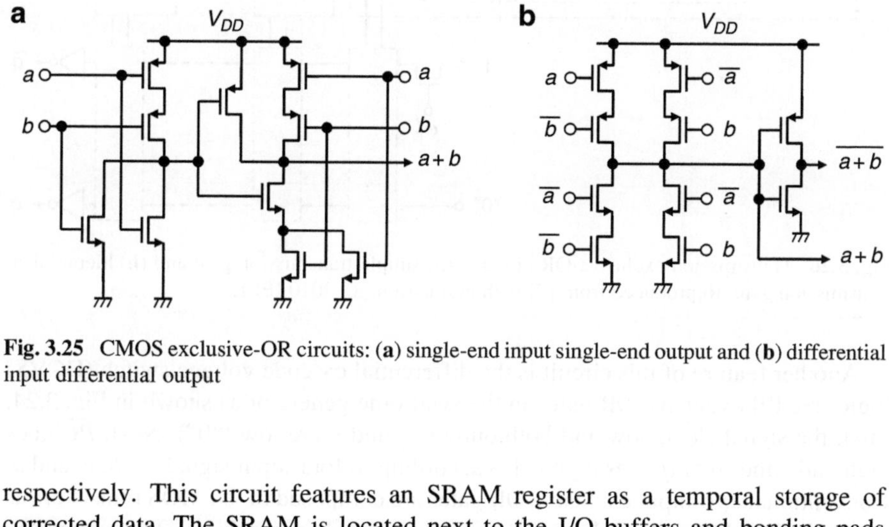

Fig. 3.25 CMOS exclusive-OR circuits: (**a**) single-end input single-end output and (**b**) differential input differential output

respectively. This circuit features an SRAM register as a temporal storage of corrected data. The SRAM is located next to the I/O buffers and bonding pads (not shown in the figure), realizing high data rate. During read operation, 9-bit syndrome is generated from the data read from (128 + 9) memory cells and is sent to the syndrome bus. The data bits are corrected by the correction circuit using the syndrome and are written into the SRAM, while the check bits are not corrected. During write operation, the data stored in the SRAM are written into the DRAM and check bits are generated. The syndrome generator is used as a coding circuit as shown in Fig. 3.8b. The generated check bits are written into nine DRAM cells through the syndrome bus and the switch (connected to “Normal”). During test operation, the data stored in the nine SRAM cells are directly written into the nine DRAM cells by connecting the switch to “Test”.

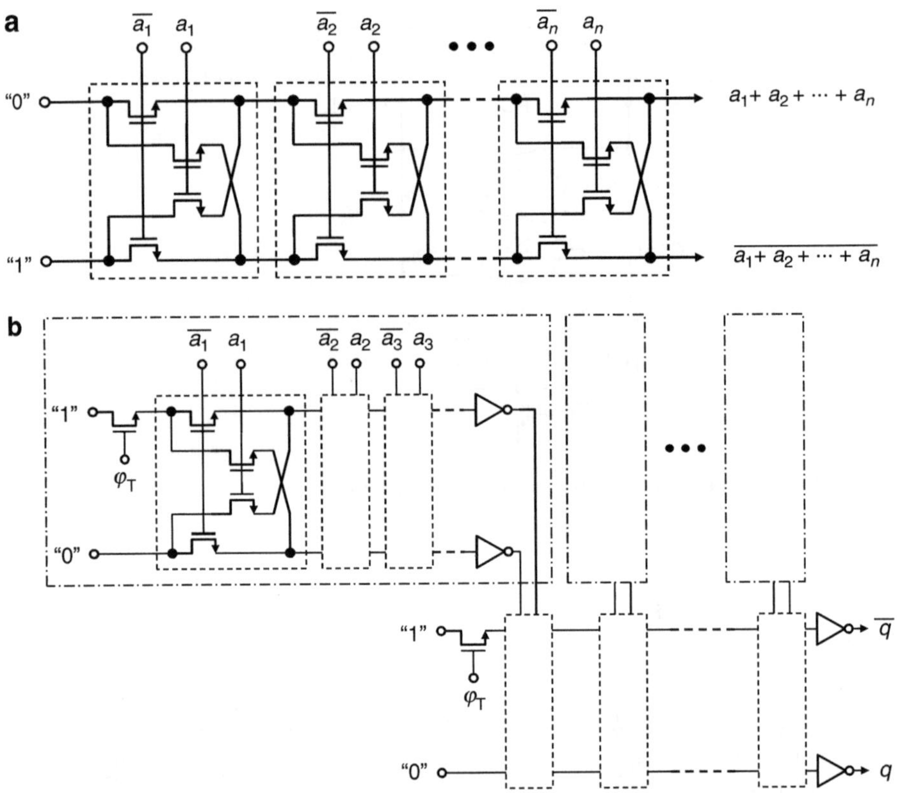

Fig. 3.26 Transmission exclusive-OR circuits: (**a**) simple transmission gate and (**b**) hierarchical transmission gate. Reproduced from [7] with permission; © 2010 IEEE

Another feature of this circuit is the differential cascode voltage switch (DCVS) logic-tree [9] exclusive OR gates in the syndrome generator as shown in Fig. 3.24. First, the signal $\overline{PC}$ is low and both outputs q and $\bar{q}$ are low ("0"). Next, $\overline{PC}$ goes high and either q or $\bar{q}$ goes high ("1") according to four input signals a, b, c, and d. Conventional two-input exclusive OR gates are composed of ten MOS transistors as shown in Fig. 3.25a, b, while the four-input circuit shown in Fig. 3.24 is composed of 23 transistors. Thus, MOS transistor count of the syndrome generator is reduced by 20%. In addition the number of stages is halved.

Since exclusive OR circuits are the major portion of a coding/decoding circuit as described in Sect. 3.5, their area and delay time are important design factors. Figure 3.26a shows the transmission-type exclusive-OR circuit used for a 16-Mbit DRAM [10]. This circuit requires only four MOS transistors for two-input exclusive OR and is located next to a memory array to shorten input signal lines. Figure 3.26b shows a hierarchical transmission-type circuit for further reduction in delay time. It is applied to the above-described ECC circuit using the bidirectional parity code [7].

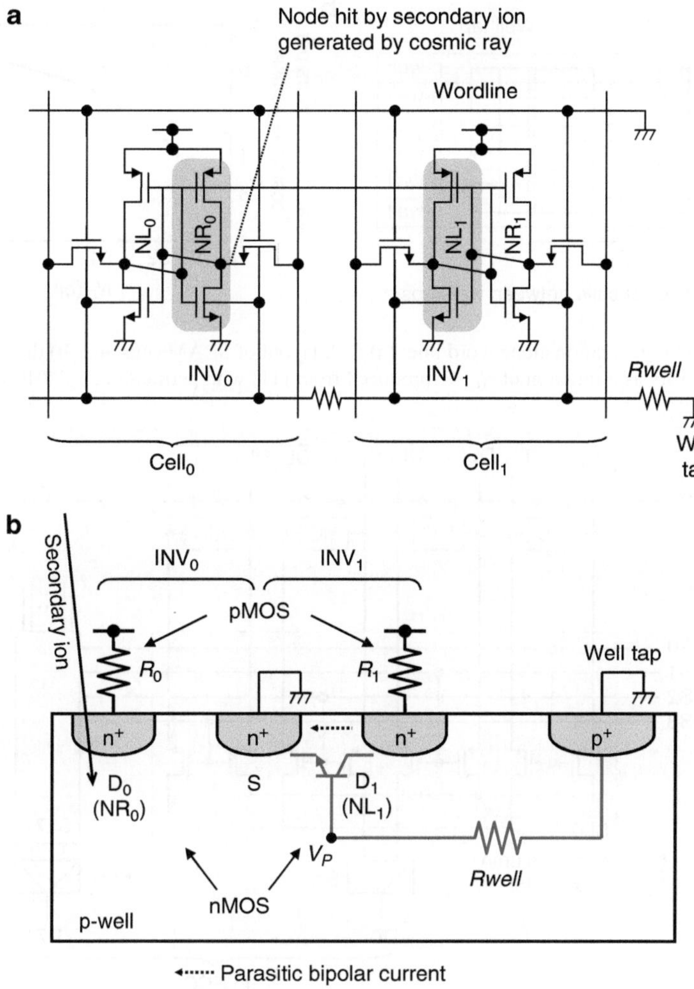

Fig. 3.27 Models used in circuit simulation (**a**) and device simulation (**b**). Reproduced from [11] with permission; © 2010 IEEE

3.7.1.3 Multicell Error Problem

In the case of SEC code, we must pay attention to the possibility of multicell error caused by a single incident of particle. Multicell errors caused by cosmic rays are intensively studied especially for SRAMs. Figure 3.27 shows the models [11] used in circuit and device simulations for two adjacent memory cells (Cell_0, Cell_1). Nodes NR_0 and NR_1 are initially at a high voltage. In the device-simulation model (Fig. 3.27b), the pMOS transistors in inverters are replaced by resistance (R_0, R_1) and the nMOS transistors are replaced by n^+ diffusion areas (D_0, S, and D_1). Substrate nodes of nMOS transistors are tied together and connected to a parasitic

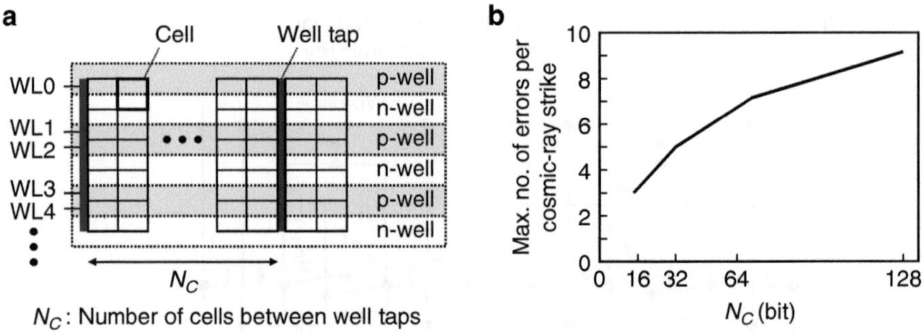

Fig. 3.28 Error generation along word line: (**a**) well layout of SRAM cells and (**b**) the maximum number of errors as a function of N_C. Reproduced from [11] with permission; © 2010 IEEE

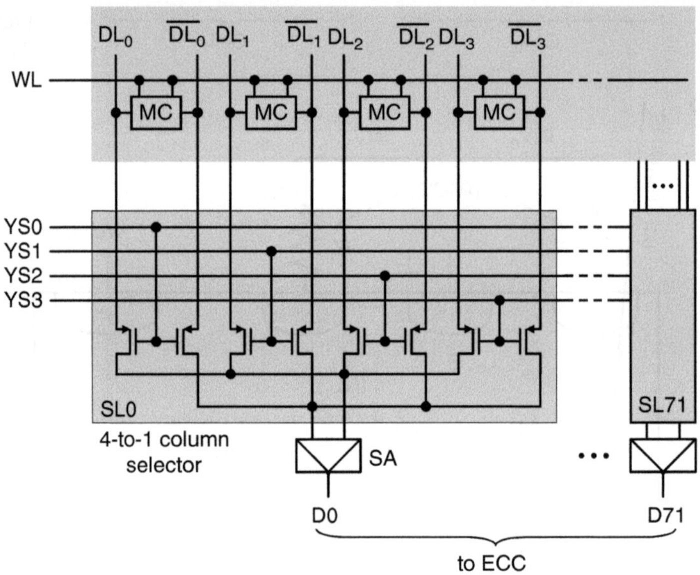

Fig. 3.29 Column selectors applied to read path [11]

p-well resistance (R_{well}), which is connected to a well tap. A parasitic bipolar device between D_1 and S is formed. The simulation results show that soft errors may occur both in Cell_0 and in Cell_1 by a single incident of ion at NR_0. The mechanism can be explained as follows [11]. After the ion hits NR_0, the funneling effect [12] leads to the very rapid collection of electrons in NR_0. On the other hand, the holes generated remain in the p-well. The p-well voltage (V_p) thus floats up to 0.9 V. This V_p switches the parasitic bipolar element on, which leads to a current flow from node NL_1 to S. Thus, NL_1 voltage slowly falls. The maximum number of multicell errors has been reported to depend on the number (N_C) of cells between well taps in the wordline direction as shown in Fig. 3.28b. This assumes $R_{\text{well}} = 1\ \text{k}\Omega/\text{cell}$ and

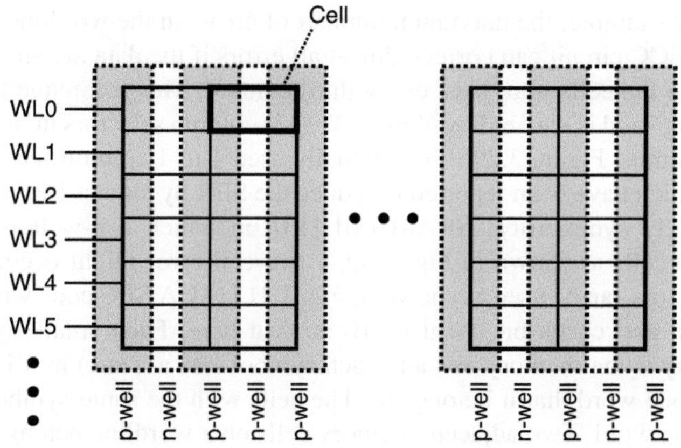

Fig. 3.30 Well layout of lithographically symmetrical SRAM cells [13]

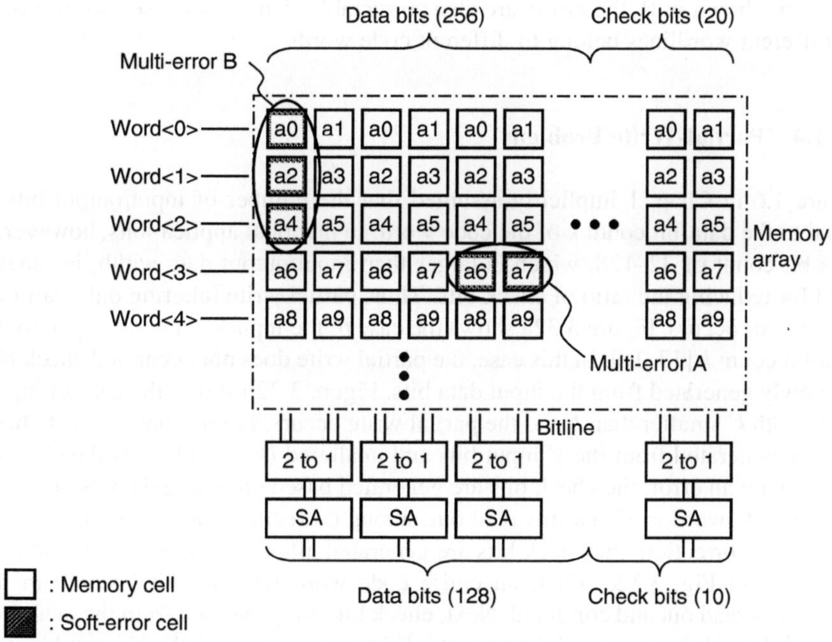

Fig. 3.31 Column selectors and physical address arrangement for the handling of multi-cell errors. Reproduced from [14] with permission; © 2010 IEEE

$C_p = 1$ fF/cell. The p-wells and n-wells run along the wordlines (Fig. 3.28a). For $N_C = 16$, three errors occur.

The key to correcting multicell errors through ECC is to avoid simultaneous reading out from multiple cells in which a multicell error might occur. For

$N_C = 16$, for example, the maximum number of errors in the wordline direction is three, so an ECC circuit can correct almost all errors if the data are simultaneously read out from the cells in at least every third column. This technique [8] is called "interweaving" and is realized by N-to-1 ($N \geq 3$) column selectors in order to select one of N columns. Figure 3.29 shows actually used 4-to-1 column selectors for the read path, which have been reported to reduce the SER by almost 100% [11]. In the lithographically symmetrical SRAM cell [13], in which a p-well is shared by adjacent two cells as shown in Fig. 3.30, a two-cell error might occur. So 2-to-1 column selectors can be used as shown in Fig. 3.31 [14]. A SEC code with a data-bit count of 128 and check-bit count of 10 is used here. Each small square in the memory array is the memory cell and each symbol (a0, a1, etc.) in it indicates the respective code word that it belongs to. The cells with the same symbol belong to the same code word. Two adjacent memory cells on a wordline belong to different code word. Even if a single incident of alpha particle or neutron causes soft errors on two adjacent cells as shown in A, each code word (a6, a7) has a single-bit error, which can be corrected. If a single incident causes soft errors on multiple cells on a bitline as shown in B, the errors are also correctable. This is because memory cells on different wordlines belong to different code words.

3.7.1.4 Partial-Write Problem

Figure 1.6 in Chap. 1 implicitly assumed that the number of input/output bits is equal to the data-bit count k of the code word. In practical applications, however, a data-bit count of 32–128, which is larger than input/output data width, is usually used for reducing the ratio of check bits. Thus, partial write (altering only part of a code word) occurs. Figure 3.32a shows the case of the input-data width equal to the data-bit count k [17, 18]. In this case, the partial write does not occur and check bits are newly generated from the input data bits. Figure 3.32b shows the case of input-data width k' smaller than k and the partial write occurs. Let us consider that check bits are generated from the k' input bits and unaltered $(k - k')$ bits. If the $(k - k')$ bits contain an error, the check bits are generated based on wrong data. When (part of) the code word is afterwards read out, wrong correction may occur. To prevent the wrong correction, the check bits are generated after the correction of read data as shown in Fig. 3.32c. First, an entire code word [$(k + m)$ bits] stored in the memory is read out and corrected. Next, check bits are generated from the unaltered $(k - k')$ bits of the corrected data and the k' input bits. Finally, the k' input bits and m check bits are written into the memory. The 16-Mbit DRAM [8] described in the previous subsection adopts this scheme. The demerit of this scheme is a large access-time penalty because the critical path contains both the coding and decoding circuits. The 16-Mbit DRAM mitigates the penalty by the high-data-rate input/output of the SRAM.

Figure 3.32d shows the scheme proposed in [15]. It features that a two-port memory is used for check bits. Since the generation of check bits requires a considerable time as mentioned above, the check bits are written into the

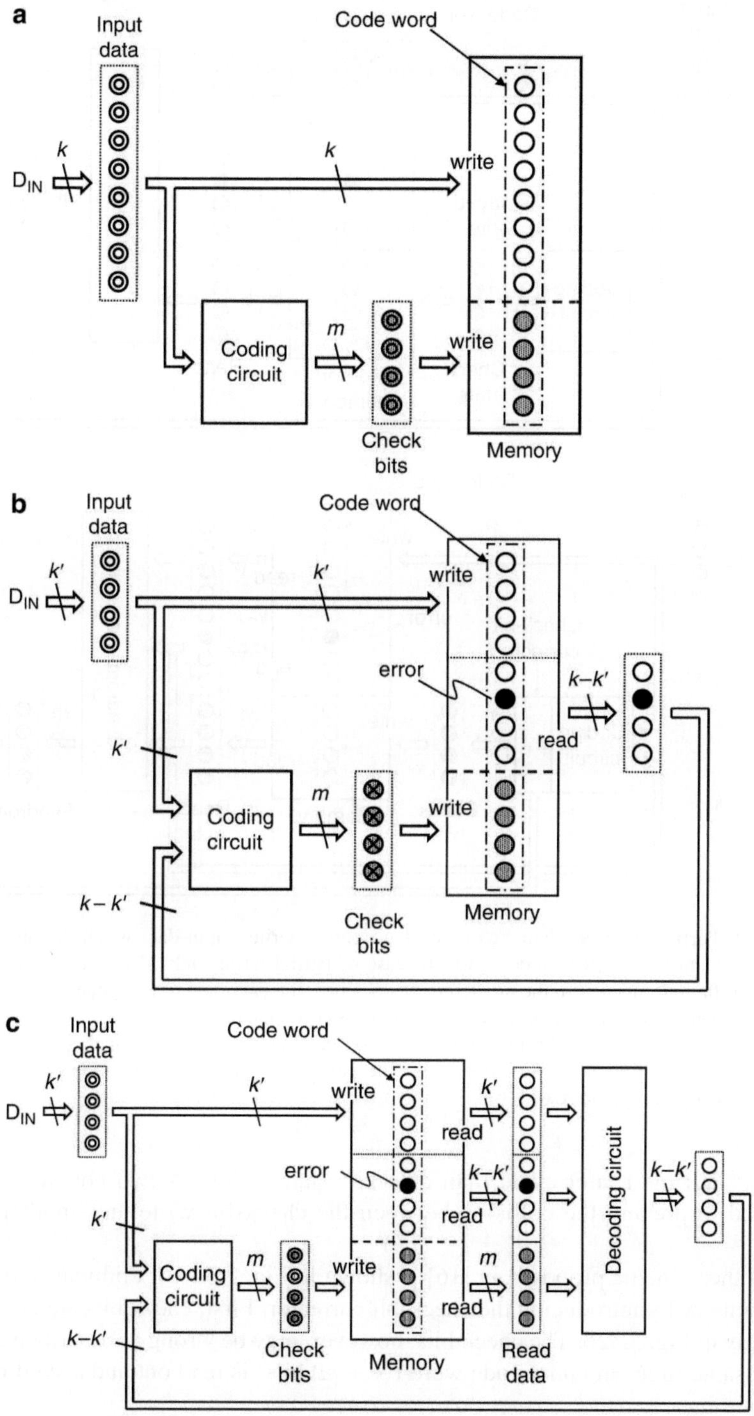

Fig. 3.32 *continued*

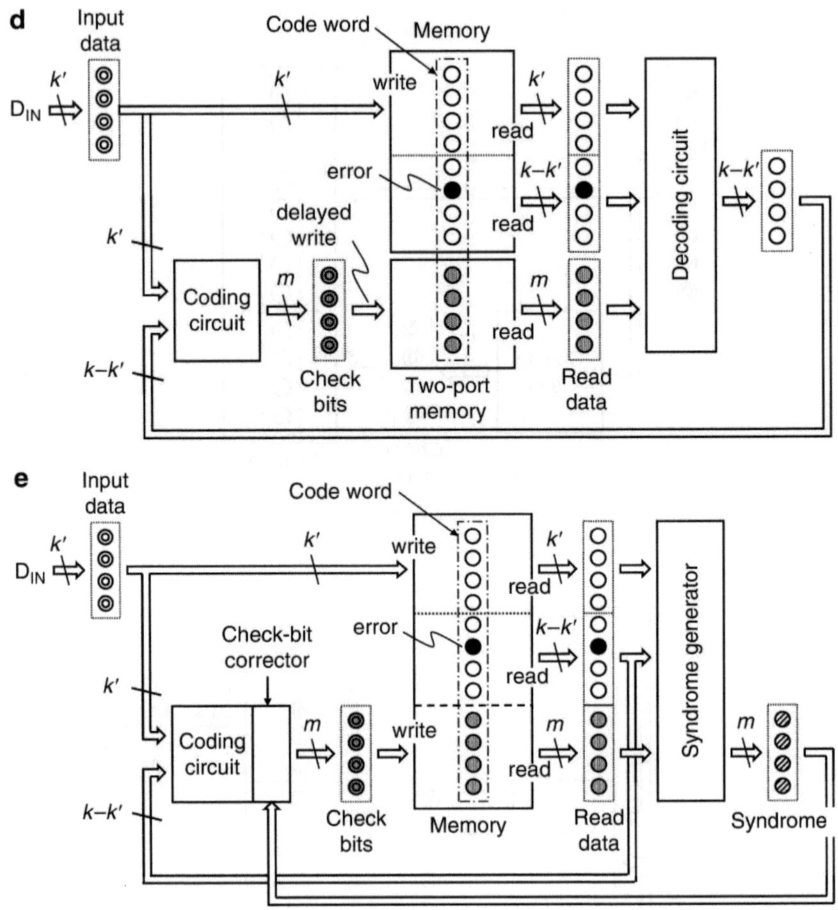

Fig. 3.32 Partial-write problem: (**a**) case of no partial write; input-data width is equal to the number of data bits of ECC code word, (**b**) case of partial write (only k' bits among k bits are altered); if there is an error in the unaltered $(k - k')$ bits, the encoding circuit generates parity bits based on wrong data, (**c**) an error in the $(k - k')$ bits is corrected by the decoding circuit; the access penalty is large because both the decoding and encoding circuits are in the critical path, (**d**) access-time-penalty reduction using two-port cells for parity bits [15], and (**e**) access-time-penalty reduction using parity corrector [16]

memory during a later cycle than data bits (late write). A two-port memory is required to prevent the collision between the check-bit write and another read/write.

Another scheme proposed in [16] is shown in Fig. 3.32e. It eliminates the two-port memory by introducing the check-bit corrector. First, check bits are generated similarly to Fig. 3.32b. The check bits, however, may be wrong as described above. At the same time, an entire code word ($(k + m)$ bits) is read out and a syndrome is generated. By analyzing the syndrome we know whether the unaltered $(k - k')$ contains an error, and whether the generated check bits are wrong. The check bits are corrected using this information.

[Example] Let us consider the case of using a shortened Hamming code (data-bit count $k = 8$, and check-bit count $m = 4$) with a check matrix of

$$\boldsymbol{H} = \begin{pmatrix} 1 & 1 & 1 & 0 & 0 & 0 & 1 & 1 & 1 & 0 & 0 & 0 \\ 1 & 0 & 0 & 1 & 1 & 0 & 1 & 1 & 0 & 1 & 0 & 0 \\ 0 & 1 & 0 & 1 & 0 & 1 & 1 & 0 & 0 & 0 & 1 & 0 \\ 0 & 0 & 1 & 0 & 1 & 1 & 0 & 1 & 0 & 0 & 0 & 1 \end{pmatrix}, \tag{3.108}$$

and altering the first four bits. If one of the unaltered four bits, for example the sixth bit, is erroneous, the syndrome $\boldsymbol{s}$ is equal to the sixth column of $\boldsymbol{H}$.

$$\boldsymbol{s} = \begin{pmatrix} 0 \\ 0 \\ 1 \\ 1 \end{pmatrix}. \tag{3.109}$$

In this case, the last two bits of the generated check bits are erroneous and must be inverted. Generally, if $\boldsymbol{s}$ is equal to one of the fifth to eighth columns of $\boldsymbol{H}$,

$$\begin{pmatrix} 0 \\ 1 \\ 0 \\ 1 \end{pmatrix}, \begin{pmatrix} 0 \\ 0 \\ 1 \\ 1 \end{pmatrix}, \begin{pmatrix} 1 \\ 1 \\ 1 \\ 0 \end{pmatrix}, \text{or} \begin{pmatrix} 1 \\ 1 \\ 0 \\ 1 \end{pmatrix},$$

the correct check bits are given by the exclusive OR of the generated check bits and $\boldsymbol{s}$.

3.7.1.5 Startup Problem

Since the data in volatile memories are lost by power-off, the data stored in them just after power-on are undefined. This means that the data stored in volatile memories with on-chip ECC may contain errors. In particular, check bits are generated based on partially undefined data in the case of partial write described above. A solution to this problem is writing error-free data into all the memory cells after power-on [19]. The simplest way is clearing all the cells using, for example, the methods proposed in [20] or [21].

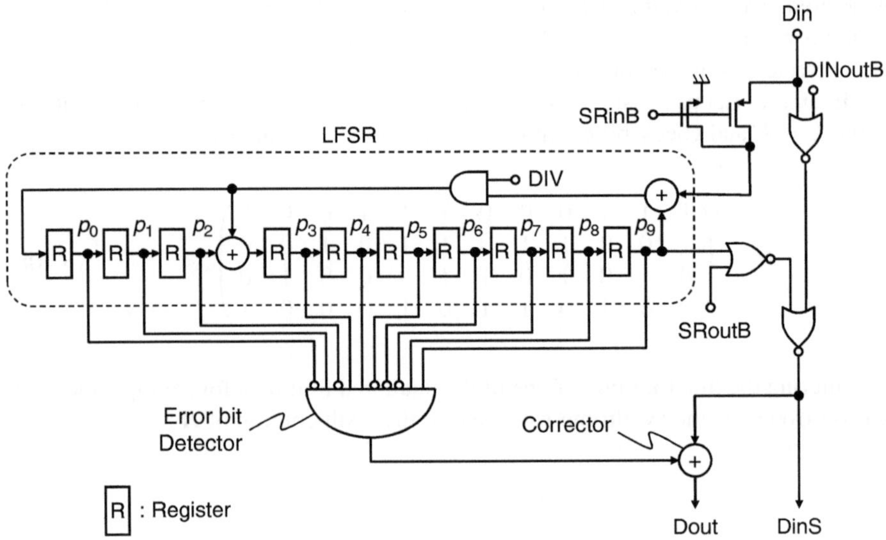

Fig. 3.33 Serial ECC circuit. Reproduced from [22] with permission; © 2010 IEEE

3.7.2 Application to Serial-Access Memories

The serial coding/decoding scheme by an LFSR using a cyclic code described in Sect. 3.5 is effective for the on-chip ECC of serial memories where data are inputted or outputted in synchronous with a clock signal. Figure 3.33 shows the coding/decoding circuit for a 64-Mbit NAND flash memory [22]. The error correcting code used here is a cyclic Hamming code, whose generator polynomial is expressed as

$$G(x) = x^{10} + x^3 + 1. \tag{3.110}$$

Since $x^{1023} + 1$ can be divided by $G(x)$, the code length, check-bit count, and data-bit count of the cyclic code are 1023, 10, and 1013, respectively. The code is shortened by 501 bits to obtain a shortened code with a code length of 522 and a data-bit count of 512. This circuit features that an LFSR is common to coding and decoding. The waveforms of write and read operations are shown in Fig. 3.34a, b, respectively.

The coding procedure utilizes the method shown in Fig. 3.11b. First, the signals DINoutB and SRinB are low and DIV and SRoutB are high. Input data bits d_{501}, d_{502}, …, d_{1012} are serially inputted from the Din terminal (d_0, d_1, …, d_{500} are assumed as zero because of shortened code). The input data bits are sent to the output terminal DinS as output data bits w_{501}, …, w_{1012} as they are, while check bits

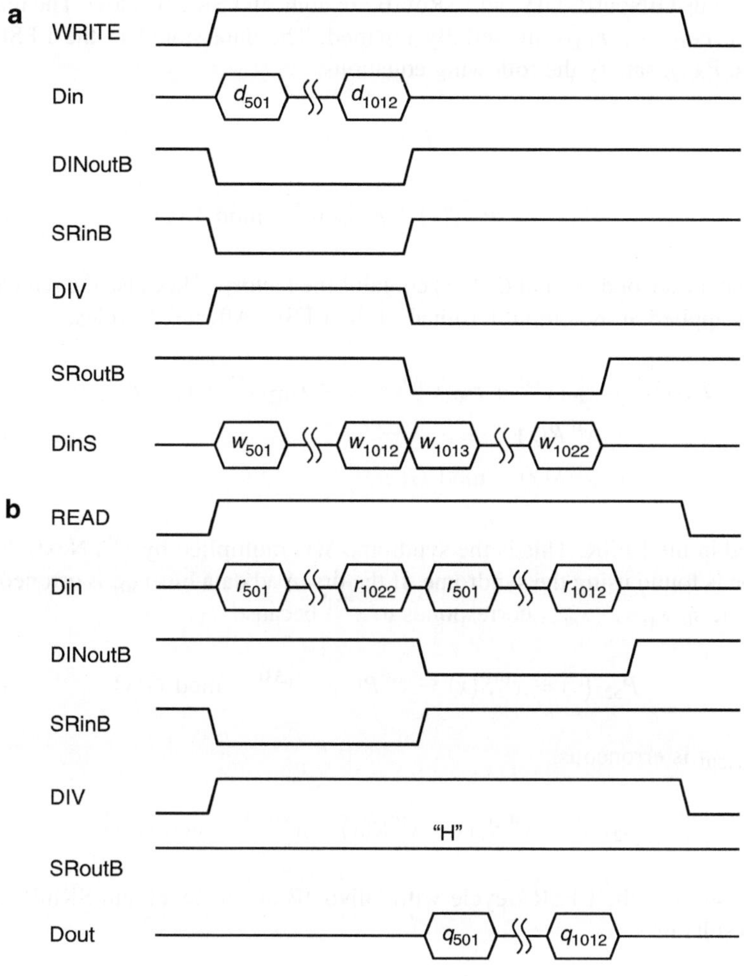

Fig. 3.34 Timing chart of ECC circuit: (**a**) write mode and (**b**) read mode. Reproduced from [22] with permission; © 2010 IEEE

are calculated by the LFSR. After 512 cycles the check bits are stored in the LFSR. Next, signals DINoutB and SRinB go high and DIV and SRoutB go low. By applying clock signals, the check bits $w_{1013}, \ldots, w_{1022}$ stored in the LFSR are serially outputted to the DinS terminal.

The decoding procedure is different from Fig. 3.12 in that the read data is applied to the output terminal of the LFSR. Read data bits $r_{501}, r_{502}, \ldots, r_{1022}$ (r_{501}–r_{1012}: data bits, r_{1013}–r_{1022}: check bits) are expressed by a polynomial

$$R(x) = r_{501}x^{521} + r_{502}x^{520} + \cdots + r_{1021}x + r_{1022}. \tag{3.111}$$

First, signals DINoutB, DIV, and SRoutB are high, and SRinB is low. The read data bits r_{501}, r_{502}, ..., r_{1022} are serially inputted. The data stored in the LFSR after i cycles, $P_i(x)$, satisfy the following equations:

$$P_0(x) = 0, \tag{3.112}$$

$$P_i(x) = xP_{i-1}(x) + r_{i+500}x^{10} \quad \text{mod } G(x). \tag{3.113}$$

Note that the second term of (3.113) contains the factor x^{10} because the read data bit r_{i+500} is applied at the output terminal of the LFSR. After 522 cycles,

$$\begin{aligned} P_{522}(x) &= r_{501}x^{531} + r_{502}x^{530} + \cdots + r_{1021}x^{11} + r_{1022}x^{10} \\ &= x^{10}R(x) \\ &= x^{10}S(x) \quad \text{mod } G(x), \end{aligned} \tag{3.114}$$

is stored in the LFSR. This is the syndrome $S(x)$ multiplied by x^{10}. Next, the error position is found using the syndrome. If the first read data bits r_{501} is erroneous, the syndrome s_0, s_1, ..., s_{m-1} corresponds to x^{531} because

$$P_{522}(x) = x^{10}S(x) = x^{10}R(x) = x^{531} \quad \text{mod } G(x). \tag{3.115}$$

If bit r_{i+501} is erroneous,

$$P_{522}(x) = x^{10}S(x) = x^{10}R(x) = x^{531-i} \quad \text{mod } G(x). \tag{3.116}$$

Further shifting the LFSR i cycle with DINoutB at low level and SRinB at high-level results in

$$P_{522+i}(x) = x^i \cdot x^{10}S(x) = x^i \cdot x^{10}R(x) = x^{531} \quad \text{mod } G(x). \tag{3.117}$$

Thus, the data stored in LFSR correspond to x^{531}. The correction is therefore performed by inputting the read data in synchronous with the clock and correcting the data bit when the stored data is x^{531}. Since

$$x^{531} = x^9 + x^8 + x^4 + x^3 + x \quad \text{mod } G(x), \tag{3.118}$$

the data bit is corrected when $p_0 = 0$, $p_1 = 1$, $p_2 = 0$, $p_3 = 1$, $p_4 = 1$, $p_5 = 0$, $p_6 = 0$, $p_7 = 0$, $p_8 = 1$, and $p_9 = 1$.

The hardware of the serial coding/decoding circuit is far smaller than that of the parallel circuit. It is reported that the chip-size penalty due to ECC of the 64-Mbit flash memory is as small as 1.9% [22].

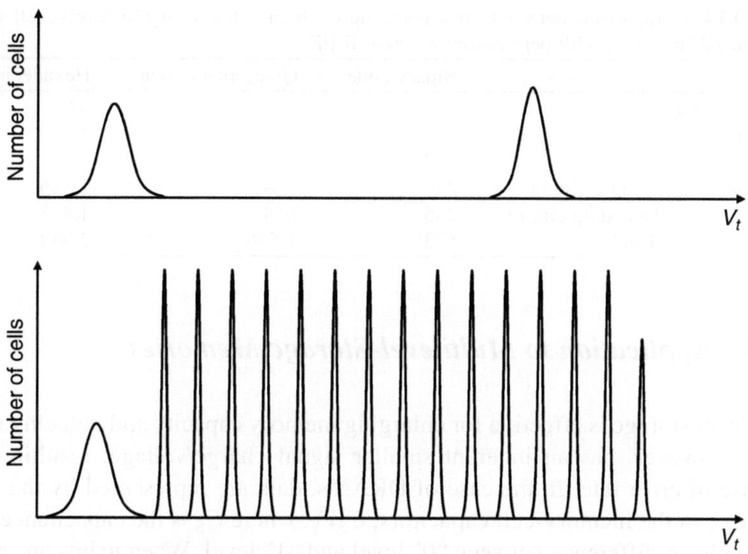

Fig. 3.35 Conceptual representation of (**a**) bilevel and (**b**) 16-level threshold voltage (V_t) distributions of flash memories. Reproduced from [23] with permission; © 2010 IEEE

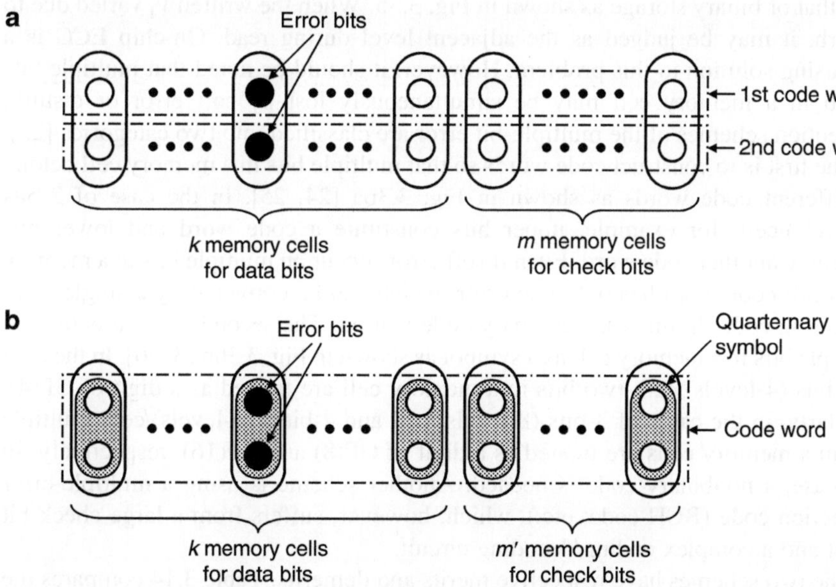

Fig. 3.36 Error correction schemes for 2-bit (4-level)/cell memory: (**a**) using a binary code; upper bit and lower bit of each cell belongs to separate code words [24, 25] and (**b**) using a quarternary code; 2 bits stored in a cell is treated as a quarterary symbol (a digit on GF(4)) [3, 26]

Table 3.14 Comparison between error-correction schemes for 4-bit (16-levels)/cell memory. Reproduced from [23] with permission; © 2010 IEEE

Code		Binary code	Quarternary code	Hexadecimal code
Number of data cells		64	64	64
Number of check cells		7	4	3
Operating modes (bit/cell)		1, 2, 3, 4	2	4
Gate count	Coding circuit	278	738	1,526
	Decoding circuit	245	618	1,423
	Total	523	1,536	2,949

3.7.3 Application to Multilevel-Storage Memories

Multilevel storage is effective for enlarging memory capacity and reducing cost per bit. It, however, has an inherent smaller signal charge/voltage, resulting in the increase of error rate. In the case of DRAMs, data are represented by the amount of charge in the memory-cell capacitors, $C_S V_S$, where C_S is the capacitance and V_S is the voltage difference between "0" level and "1" level. When m bits are stored in a memory cell, 2^m levels are required. Thus, the difference between adjacent levels is $1/(2^m - 1)$ times that of a binary memory. It means that the signal charge and the critical charge of soft error become $1/(2^m - 1)$ times. In the case of flash memories, data are represented by the threshold voltages (V_t) of cell transistors. When multiple bits are stored in a memory cell, V_t difference between adjacent levels is smaller than that of binary storage as shown in Fig. 3.35. When the written V_t varied due to disturb, it may be judged as the adjacent level during read. On-chip ECC is a promising solution to this problem. However, it should be noted that multiple bits stored in a memory cell may be simultaneously lost by soft error or disturb. Correction schemes of the multiple-bit error are classified into two categories [23].

The first is to construct code words so that multiple bits in a memory cell belong to different code words as shown in Fig. 3.36a [24, 25]. In the case of 2-bits (4-levels)/cell, for example, upper bits constitute a code word and lower bits constitute another code word. Even if soft errors occur on multiple bits in a memory cell, each code word has only one error, which can be corrected by a single-error correcting code. In this case, a binary code is used. The second scheme is to treat multiple bits in a memory cell as a symbol as shown in Fig. 3.36b [3, 26]. In the case of 2-bits (4-levels)/cell, two bits in a memory cell are treated as a digit of GF(4). Similarly, in the cases of 3-bits (8-levels)/cell and 4-bits (16-levels)/cell, multiple bits in a memory cells are treated as a digit of GF(8) and GF(16), respectively. In this case, a nonbinary code is needed. Another scheme is using a multiple-error correction code (BCH code, etc.), which, however, suffers from a large check-bit count and a complex coding/decoding circuit.

The two schemes have respective merits and demerits. Table 3.14 compares the first scheme (binary code), the second scheme (hexadecimal code) and an intermediate scheme (quaternary code) for the case of 4-bits (16-levels)/cell storage [23]. The second scheme requires fewer memory cells for check bits, as described in

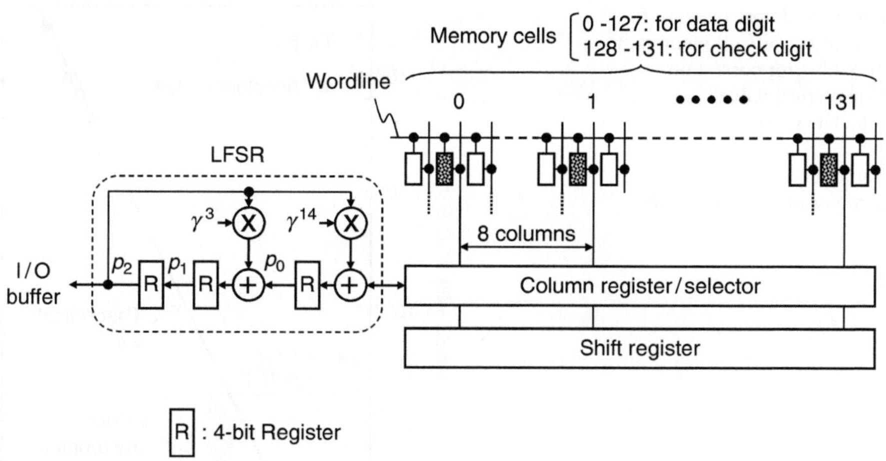

Fig. 3.37 Serial error correction circuit for a 4-bit/cell memory. *Shaded cells* construct a code word. Reproduced from [3] with permission; © 2010 IEEE

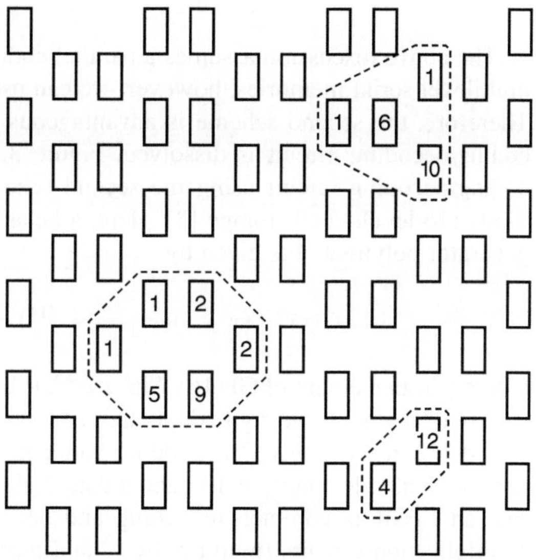

Fig. 3.38 Typical cell soft-error patterns of a 16-lelvel dynamic memory caused by incidents of alpha particles. The *rectangles* are the locations of trench capacitors in the memory array. Each *dashed line* indicates a group of memory cells in which soft error occurred by a single incident. Each *digit* indicates the error level of each cell. Reproduced from [3] with permission; © 2010 IEEE

Sect. 3.4 (Table 3.10). However, it requires a larger coding/decoding circuit. The selection of the scheme, therefore, requires the trade-off between check-bit count and hardware size. The first scheme has another merit that the on-chip ECC is available in whichever mode (1 bit/cell, 2 bits/cell, 3 bits/cell, or 4 bits/cell) the memory operates.

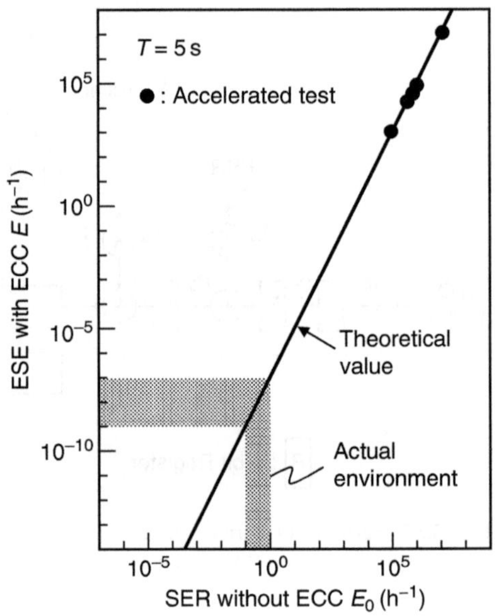

Fig. 3.39 Improvement of SER of a 4-bit/cell memory through error correction. "Theoretical value" is calculated from (3.84) ($W = 8{,}192$, $T = 5$ s). Reproduced from [3] with permission; © 2010 IEEE

The above discussion assumes a parallel coding/decoding circuit. In the case of multilevel serial memories, however, we can use a serial coding/decoding circuit. Therefore, the second scheme is advantageous because the demerit of the larger coding/decoding circuit is dissolved. Figure 3.37 shows the schematic of serial coding/decoding circuit using the second scheme for a dynamic memory using 4-bits (16-levels)/cell storage [3]. Here, a hexadecimal cyclic code is used whose generator polynomial is given by

$$G(x) = (x + 1)(x^2 + x + \gamma^{14}) = x^3 + \gamma^3 x + \gamma^{14}, \tag{3.119}$$

where γ is an element of GF(16). Since $x^{255} + 1$ on GF(16) can be divided by $G(x)$, the code length, check-digit count, and data-digit count of the cyclic code are 255, 3, and 252, respectively. The code is shortened by 124 digits to obtain a shortened code with a code length of 131 and a data-digit count of 128. This circuit features that an LFSR is common to coding and decoding as in the case of Fig. 3.33. Multiplication circuits (multiply-by-γ^3 and multiply-by-γ^{14} circuits) are provided in the LFSR because of the nonbinary code. The interweaving technique is used as in the case of Fig. 3.31 to cope with multicell errors, the examples of which are shown in Fig. 3.38. Memory cells of every eighth column belong to the same code word. It is reported that the results of soft-error acceleration tests using various alpha-particle sources are close to the theoretical value of (3.84) as shown in Fig. 3.39.

In the case of multilevel serial memories, multiple-error correcting codes are also attractive because serial coding/decoding circuits can be used. It is reported

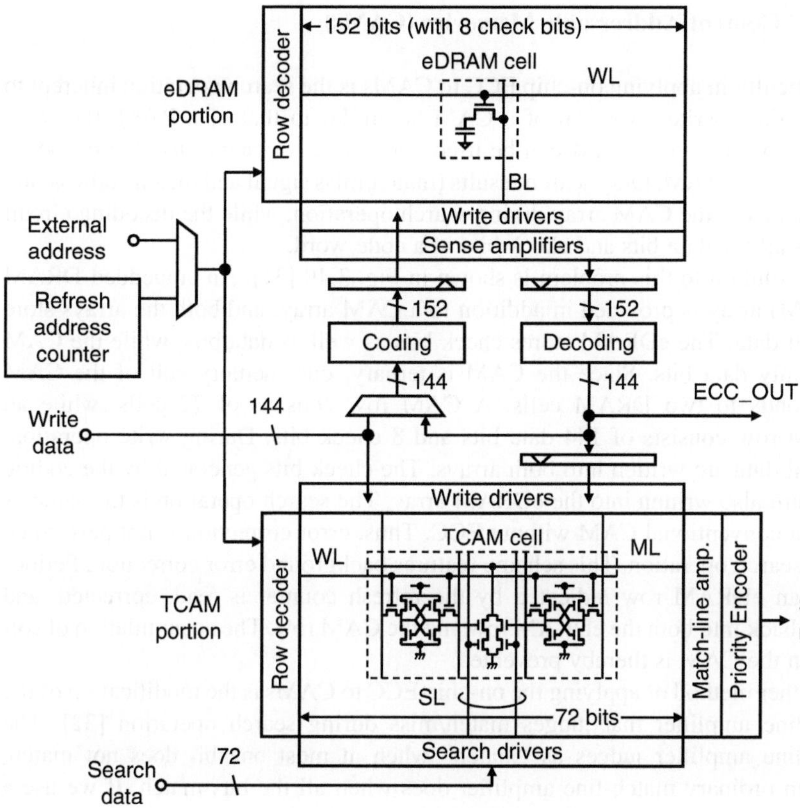

Fig. 3.40 Block diagram of TCAM with error correction. Reproduced from [31] with permission; © 2010 IEEE

that the chip-size penalty of a 4-Gbit 2-bits/cell NAND flash memory using a quintuple-error correcting BCH code is less than 1% [27].

3.7.4 Application to Other Memories

3.7.4.1 Mask ROMs

It is difficult to apply conventional redundancy techniques to mask ROMs. Not only faulty addresses but also the data stored at the addresses must be programmed during fabrication. Thus, nonvolatile devices for storing the data are required [28]. On-chip ECC is effective for the hard errors of mask ROMs [29, 30]. The coding circuit is not needed because of read only. The partial-write and startup problems described above do not occur.

3.7.4.2 Content Addressable Memories (CAMs)

The difficulty in applying on-chip ECC to CAMs is the search operation inherent to CAMs. Since write operation of a CAM is similar to that of a RAM, the same coding procedure as a RAM can be used. However, we cannot use the decoding procedure of a RAM. Only search results (match/miss signal and match address) are outputted from the CAM array during search operation, while the decoding circuit requires all the data bits and check bits of a code word.

One solution to this problem is shown in Fig. 3.40 [31]. An embedded-DRAM (eDRAM) array is provided in addition to a CAM array, and both the arrays store identical data. The eDRAM stores check bits as well as data bits, while the CAM stores only data bits. Since the CAM is ternary, one memory cell of the CAM corresponds to two DRAM cells. A CAM row consists of 72 cells, while an eDRAM row consists of 144 data bits and 8 check bits. During write operation, identical data are written into both arrays. The check bits generated by the coding circuit are also written into the eDRAM array. The search operation is the same as that of a conventional CAM without ECC. Thus, error correction is not performed during search operation. This scheme features background error correction. Periodically, an eDRAM row indicated by the refresh counter is read, corrected, and written back into both the eDRAM row and the CAM row. The accumulation of soft errors in the CAM is thereby prevented.

Another method of applying the on-chip ECC to CAMs is the modification of the match-line amplifier that judges match/miss during search operation [32]. The match-line amplifier judges as "match" when at most one bit does not match, while an ordinary match-line amplifier does when all the bits match. If we use a

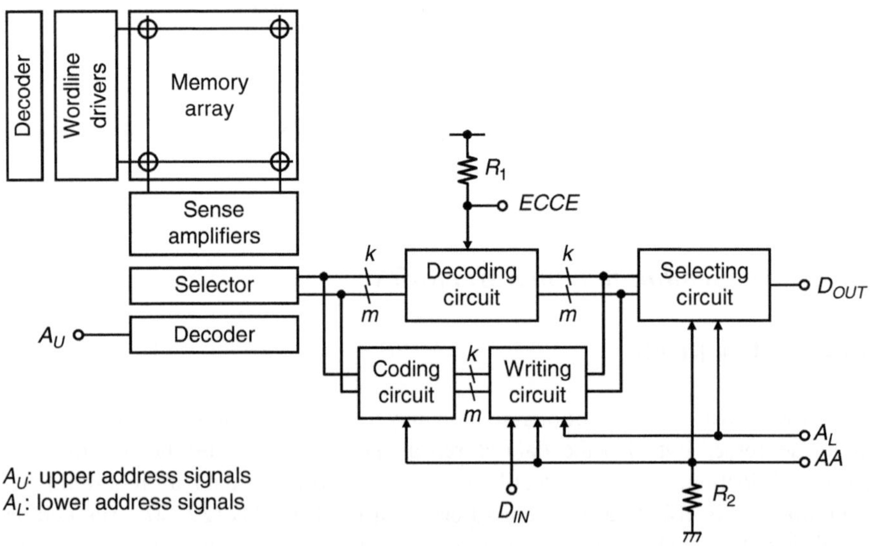

Fig. 3.41 Testing circuit for ECC [33]

code with a minimum distance of three or more, correct search results are outputted even if there is an error.

3.8 Testing for ECC

It is desirable to provide a testing circuit so that LSI vendors can test the functions of on-chip ECC circuits of memory LSIs. The testing includes the followings [18]:

1. Testing of memory cells for both data and check bits: whether each memory cell is written and read correctly.
2. Testing of coding: whether correct check bits are generated.
3. Testing of decoding: whether data with error(s) are detected and corrected.

The circuit shown in Fig. 3.41 enables the above three kinds of tests [33]. Two test signals ECCE (ECC enable) and AA (additional address) are applied from the test terminals, which are open during normal read/write operation to set ECCE and AA at high level and low level, respectively. The k data bits and m check bits selected by the selector among the data read out from the memory array are sent to the decoding circuit for error correction. During read operation, part of the k data bits is selected and is outputted through the D_{OUT} terminal(s). During write operation, part of the k data bits are replaced with the data inputted from the D_{IN} terminal(s) (partial write, see Fig. 3.32c), and are written into the memory array with m check bits generated by the coding circuit.

1. Testing of memory cells
   Signal ECCE is set at low level to disable the error correction of the decoding circuit. Memory cells are tested by successive write and read. Part of k data bits are replaced with the input bits if AA is low, and part of m check bits are replaced with the input bits if AA is high. Part of k data bits are outputted to D_{OUT} terminal if AA is low, and part of m check bits are outputted if AA is high. Thus, signal AA works as an extra address signal to select either data bits or check bits.
2. Testing of coding
   First, test data are written by normal write operation. The data bits and the check bits generated by the coding circuit are thereby written into the respective memory cells. Next, ECCE and AA are set at low and high, respectively, and the check bits are read out without error correction. Thus, we can know whether correct check bits were generated.
3. Testing of decoding
   First, test data with error(s) are written into the memory array. The data bits are written with AA at low level, and the check bits are written with AA at high level. Next, the data are read out with ECCE and AA at high and low, respectively. We can know whether the data bits are corrected. We can also know whether the check bits are corrected by the read operation with both ECCE and AA at high level.

References

1. H. Miyagawa, Y. Iwadare and H. Imai, *Code Theory*, Shokodo, 1976 (in Japanese).
2. M. Y. Hsiao, "A class of optimal minimum odd-weight-column SEC-DED codes," IBM J. Res. Dev., vol. 14, pp. 395–401, July 1970.
3. M. Horiguchi, M. Aoki, Y. Nakagome, S. Ikenaga and K. Shimohigashi, "An experimental large-capacity semiconductor file memory using 16-levels/cell storage," IEEE J. Solid-State Circuits, vol. 23, pp. 27–33, Feb. 1988.
4. K. Itoh, M. Horiguchi and T. Kawahara, "Ultra-low voltage nano-scale embedded RAMs," in Proc. ISCAS, May 2006, pp. 25–28.
5. T. Mano, J. Yamada, J. Inoue and S. Nakajima, "Circuit techniques for a VLSI memory," IEEE J. Solid-State Circuits, vol. SC-18, pp. 463–470, Oct. 1983.
6. J. Yamada, T. Mano, J. Inoue, S. Nakajima and T. Matsuda, "A submicron 1 Mbit dynamic RAM with a 4b-at-a-time built-in ECC circuit," IEEE J. Solid-State Circuits, vol. SC-19, pp. 627–633, Oct. 1984.
7. J. Yamada, "Selector-line merged built-in ECC technique for DRAM's," IEEE J. Solid-State Circuits, vol. SC-22, pp. 868–873, Oct. 1987.
8. H. L. Kalter, C. H. Stapper, J. E. Barth Jr., J. DiLorenzo, C. E. Drake, J. A. Fifield, G. A. Kelley Jr., S. C. Lewis, W. B. van der Hoeven and J. A. Yankosky, "A 50-ns 16-Mb DRAM with a 10-ns data rate and on-chip ECC," IEEE J. Solid-State Circuits, vol. 25, pp. 1118–1128, Oct. 1990.
9. L. G. Heller, W. R. Griffin, J. W. Davis and N. G. Thoma, "Cascade voltage switch logic: a differential CMOS logic family," in ISSCC Dig. Tech. Papers, Feb. 1984, pp. 16–17.
10. K. Arimoto, Y. Masuda, K. Furutani, M. Tsukude, T. Ooishi, K. Mashiko and K. Fujishima, "A speed-enhanced DRAM array architecture with embedded ECC," IEEE J. Solid-State Circuits, vol. 25, pp. 11–17, Feb. 1990.
11. K. Osada, K. Yamaguchi, Y. Saitoh and T. Kawahara, "SRAM immunity to cosmic-ray-induced multierrors based on analysis of an induced parasitic bipolar effect," IEEE J. Solid-State Circuits, vol. 39, pp. 827–833, May 2004.
12. C. M. Hsieh, P. C. Murley and R. R. O'Brien, "A field-funneling effect on the collection of alpha-particle-generated carriers in silicon devices," IEEE Electron Device Lett., vol. EDL-2, pp. 103–105, Apr. 1981.
13. K. Osada, J. L. Shin, M. Khan, Y. Liou, K. Wang, K. Shoji, K. Kuroda, S. Ikeda and K. Ishibashi, "Universal-Vdd 0.65-2.0-V 32-kB cache using a voltage-adapted timing-generation scheme and a lithographically-symmetrical cell," IEEE J. Solid-State Circuits, vol. 36, pp. 1738–1744, Nov. 2001.
14. K. Osada, Y. Saito, E. Ibe and K. Ishibashi, "16.7-fA/cell tunnel-leakage-suppressed 16-Mb SRAM for handling cosmic-ray-induced multierrors," IEEE J. Solid-State Circuits, vol. 38, pp. 1952–1957, Nov. 2003.
15. K. Kushida, N. Otsuka, O. Hirabayashi and Y. Takeyama, "DFT techniques for memory macro with built-in ECC," in Proc. MTDT, Aug. 2005, pp. 109–114.
16. P.-Y. Chen, Y.-T. Yeh, C.-H. Chen, J.-C. Yeh, C.-W. Wu, J.-S. Lee and Y.-C. Lin, "An enhanced EDAC methodology for low power PSRAM," in Proc. ITC, Oct. 2006, 30.3.
17. S. Mehrotra, T.-C. Wu, T.-L. Chiu and G. Perlegos, "A 64Kb CMOS EEROM with on-chip ECC," in ISSCC Dig. Tech. Papers, Feb. 1984, pp. 142–143.
18. J. Y. Do, J. K. Kim, H. G. Lee, J. D. Choi and H. K. Lim, "A 256k CMOS EEPROM with enhanced reliability and testability," in Symp. VLSI Circuits Dig. Tech. Papers, Aug. 1988, pp. 83–84.
19. F. I. Osman, "Error-correction technique for random-access memories," IEEE J. Solid-State Circuits, vol. SC-17, pp. 877–881, Oct. 1982.
20. J. Etoh, K. Itoh, M. Aoki and R. Hori, "Semiconductor memory capable of high-speed data erasing," US Patent, No. 4,965,769, Aug. 1997.

21. M. Horiguchi and K. Iwasaki, "Semiconductor memory," Japanese Patent, No. 2,679,981, Aug. 1997.
22. T. Tanzawa, T. Tanaka, K. Takeuchi, R. Shirota, S. Aritome, H. Watanabe, G. Hemink, K. Shimizu, S. Sato, Y. Takeuchi and K. Ohuchi, "A compact on-chip ECC for low cost flash memories," IEEE J. Solid-State Circuits, vol. 32, pp. 662–669, May 1997.
23. S. Gregori, A. Cabrini, O. Khouri and G. Torelli, "On-chip error correcting techniques for new-generation flash memories," Proc. IEEE, vol. 91, pp. 602–616, Apr. 2003.
24. S. Gregori, O. Khouri, R. Micheloni and G. Torelli, "An error control code scheme for multilevel flash memories," in Proc. International Workshop on Memory Technology, Design and Testing, Aug. 2001, pp. 45–49.
25. B. Polianskikh and Z. Zilie, "Induced error-correcting code for 2bit-per-cell multi-level DRAM," in Proc. MWSCAS, Aug. 2001, pp. 352–355.
26. D. Rossi, C. Metra and B. Ricco, "Fast and compact error correcting scheme for reliable multilevel flash memories," in Proc. MTDT, July 2002, pp. 27–31.
27. R. Micheloni, R. Ravasio, A. Marelli, E. Alice, V. Altieri, A. Bovino, L. Crippa, E. Di Martino, L. D'Onotrio, A. Gambardella, E. Grillea, G. Guerra, D. Kim, C. Missiroli, I. Motta, A. Prisco, G. Ragone, M. Romano, M. Sangalli, P. Sauro, M. Scotti and S. Won, "A 4Gb 2b/cell NAND flash memory with embedded 5b BCH ECC for 36MB/s system read throughput," in ISSCC Dig. Tech. Papers, Feb. 2006, pp. 142–143.
28. Y. Naruke, T. Iwase, M. Takizawa, K. Saito, M. Asano, H. Nishimura and T. Mochizuki, "A 16Mb mask ROM with programmable redundancy," in ISSCC Dig. Tech. Papers, Feb. 1989, pp. 128–129.
29. T. Shinoda, Y. Ohnishi, H. Kawamoto, K. Takizawa and K. Narita, "A 1Mb ROM with on-chip ECC for yield enhancement," in ISSCC Dig. Tech. Papers, Feb. 1983, pp. 158–159.
30. H. L. Davis, "A 70-ns word-wide 1-Mbit ROM with on-chip error-correction circuits," IEEE J. Solid-State Circuits, vol. SC-20, pp. 958–963, Oct. 1985.
31. H. Noda, K. Dosaka, F. Morishita and K. Arimoto, "A soft-error-immune maintenance-free TCAM architecture with associated embedded DRAM," in Proc. CICC, Sep. 2005, pp. 451–454.
32. K. Pagiamtzis, N. Azizi and F. N. Najm, "A soft-error tolerant content-addressable memory (CAM) using an error-correcting-match scheme," in Proc. CICC, Sep. 2006, pp. 301–304.
33. M. Horiguchi, M. Aoki, Y. Nakagome, S. Ikenaga and K. Shimohigashi, "Semiconductor memory having error correcting means," US Patent No. 4,726, 021, Feb. 1988.

Chapter 4
Combination of Redundancy and Error Correction

4.1 Introduction

The effects of redundancy and error checking and correction (ECC) are separately discussed in the previous chapters. However, more faults can be repaired by the combination of the redundancy and the ECC than by simply adding the effects of them. This *synergistic effect*, which was first reported in 1990 [1], is especially effective for repairing random-bit faults. Repairing many random-bit faults by redundancy is not effective. Since the replacement unit is usually a row or a column, bit faults require as many spare rows/columns except for the case of two or more bit faults located on the same row/column. On the contrary, ECC can potentially repair many random-bit faults. However, ECC using a single-error correction code can practically repair far fewer bit faults because the probability of "fault collision" (two or more faults being located in a code word) cannot be neglected as described in Sect. 3.6. By combining the redundancy and ECC, most bit faults are repaired by ECC and a few "fault collisions" are repaired by redundancy, resulting in dramatic increase of repairable faults [1–3].

Section 4.2 describes the principle of the synergistic effect and estimation of the effect using a simple model. Section 4.3 presents the repair of faults due to device mismatch as a promising application of the synergistic effect.

4.2 Repair of Bit Faults Using Synergistic Effect

4.2.1 Principle of Synergistic Effect

Figure 4.1 shows the model of the combination of redundancy and ECC using a single-error correction code. The normal memory array is composed of W code words, each of which consists of k data bits and Δk check bits (code length $n = k + \Delta k$). In addition, a spare memory array composed of R code words is provided. A code

M. Horiguchi and K. Itoh, *Nanoscale Memory Repair*, Integrated Circuits and Systems 1,
DOI 10.1007/978-1-4419-7958-2_4,

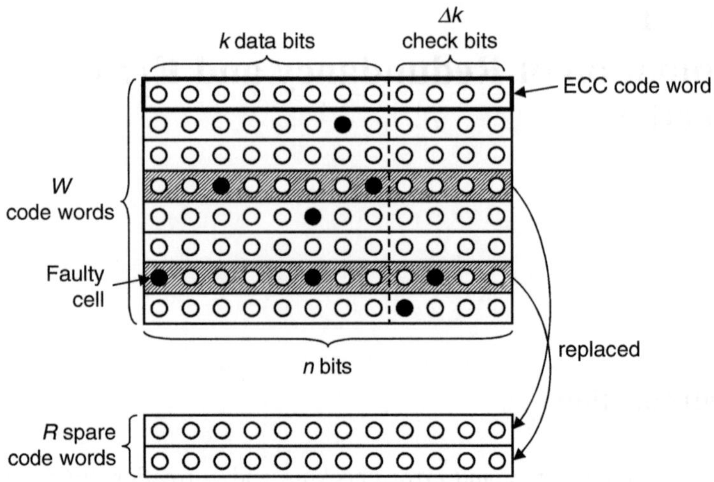

Fig. 4.1 Combination of redundancy and ECC [9]

word with only one bit fault requires no spares because the fault is corrected by ECC. A code word with two or more bit faults is replaced by a spare code word. The ECC can repair as many as W bit faults in the "lucky" case, where each code word contains a bit fault. However, the probability of "fault collision" (two or more faults being located in a code word, which cannot be corrected by ECC using a single-error correction code) is unexpectedly high as described in Sect. 3.6. Therefore, the average number of bit faults repairable by the ECC is far smaller than W. The number of "fault collisions" is usually sufficiently smaller than the number of bit faults. By replacing code words with "fault collisions" by spare code words, yield is dramatically increased. Thus, the synergistic effect is produced.

The yield calculation using this model is based on the following assumptions:

(1) The distribution of faults is random.
(2) Faults on the spare code words are neglected because R is practically far smaller than W.
(3) Faults on the check bits of ECC are not neglected because Δk is not sufficiently smaller than k.

A more practical model that is based on the real architecture of a RAM is described later.

If bit-error rate is p, the probability of a code word having no faults is $(1 - p)^n$ and the probability of a code word having one fault is $np(1 - p)^{n-1}$. Therefore, the probability of a code word being nonrepairable by ECC (having two or more faults) is expressed as

$$p_{\mathrm{cw}} = 1 - (1 - p)^n - np(1 - p)^{n-1}. \tag{4.1}$$

The probability that there are i code words nonrepairable by ECC (i.e., i spare code words are required) is expressed by binomial distribution:

$$P_i = \binom{W}{i} p_{\mathrm{cw}}^i (1 - p_{\mathrm{cw}})^{W-i}. \tag{4.2}$$

The number of code words nonrepairable by ECC must be less than or equal to R to obtain a good chip. Therefore, the yield is given by

$$Y = \sum_{i=0}^{R} P_i = \sum_{i=0}^{R} \binom{W}{i} p_{\mathrm{cw}}^i (1 - p_{\mathrm{cw}})^{W-i}. \tag{4.3}$$

In the case of ECC only (without redundancy), the yield is given by substituting $R = 0$ into (4.3):

$$Y_0 = (1 - p_{\mathrm{cw}})^W. \tag{4.4}$$

In the case of redundancy only (without ECC), a code word having one or more faults must be replaced by a spare code word. Thus, (4.1), (4.2), and (4.3) are replaced by

$$p_{\mathrm{cw}} = 1 - (1 - p)^k, \tag{4.5}$$

$$P_i = \binom{W}{i} p_{\mathrm{cw}}^i (1 - p_{\mathrm{cw}})^{W-i}, \tag{4.6}$$

and

$$Y = \sum_{i=0}^{R} P_i = \sum_{i=0}^{R} \binom{W}{i} p_{\mathrm{cw}}^i (1 - p_{\mathrm{cw}})^{W-i}, \tag{4.7}$$

respectively. Note that n is replaced by k because check bits are omitted.

The calculation result for a 256-Mbit memory using an error-correcting code with a code length of 136 and a data-bit count of 128 is shown in Fig. 4.2. The upper limit of the number of bit faults to achieve 50% yield is 128 for redundancy only ($R = 128$), and 1,711 for ECC only. On the contrary, the limit is as large as 23,402 for the combination of redundancy ($R = 128$) and ECC. This is larger than ten times the sum of both.

The calculation of (4.3) is very cumbersome. So, let us derive a simpler approximation to estimate the synergistic effect. The average of the binomial distribution of (4.2) is expressed as

$$Wp_{\mathrm{cw}} = W\{1 - (1 - p)^n - np(1 - p)^{n-1}\}, \tag{4.8}$$

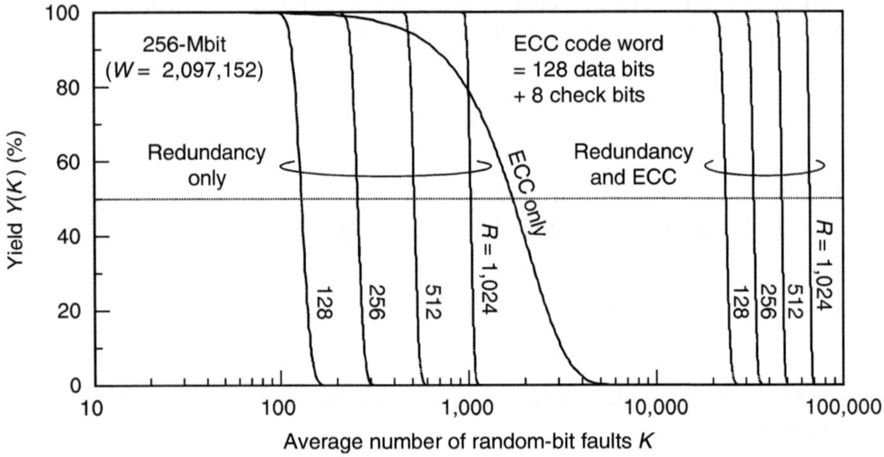

Fig. 4.2 Yield using both redundancy and ECC

which is approximated by

$$Wp_{\mathrm{cw}} \cong \frac{Wn(n-1)p^2}{2} \tag{4.9}$$

for a small p. The average number of required spare code words to obtain 50% yield is given by this equation. Since the total number of bits (including check bits) is Wn, the bit-error rate is $p = K/Wn$, where K is the total number of bit faults. Therefore

$$R \cong \frac{Wn(n-1)p^2}{2} = \frac{(n-1)K^2}{2Wn} \tag{4.10}$$

is derived. From this equation, the number of repairable bit faults is calculated as

$$K \cong \sqrt{\frac{2RWn}{n-1}}. \tag{4.11}$$

Equation (4.11) is shown by the broken lines in Fig. 4.3a, while the symbols in the figure are the precise values calculated using (4.3). The approximation of (4.11) is good for $R \geq 16$ but is worse for smaller R. Let us remind that the number of repairable bit faults without redundancy is given by

$$K_0 \cong \sqrt{\frac{2Wn \ln 2}{n-1}} \tag{4.12}$$

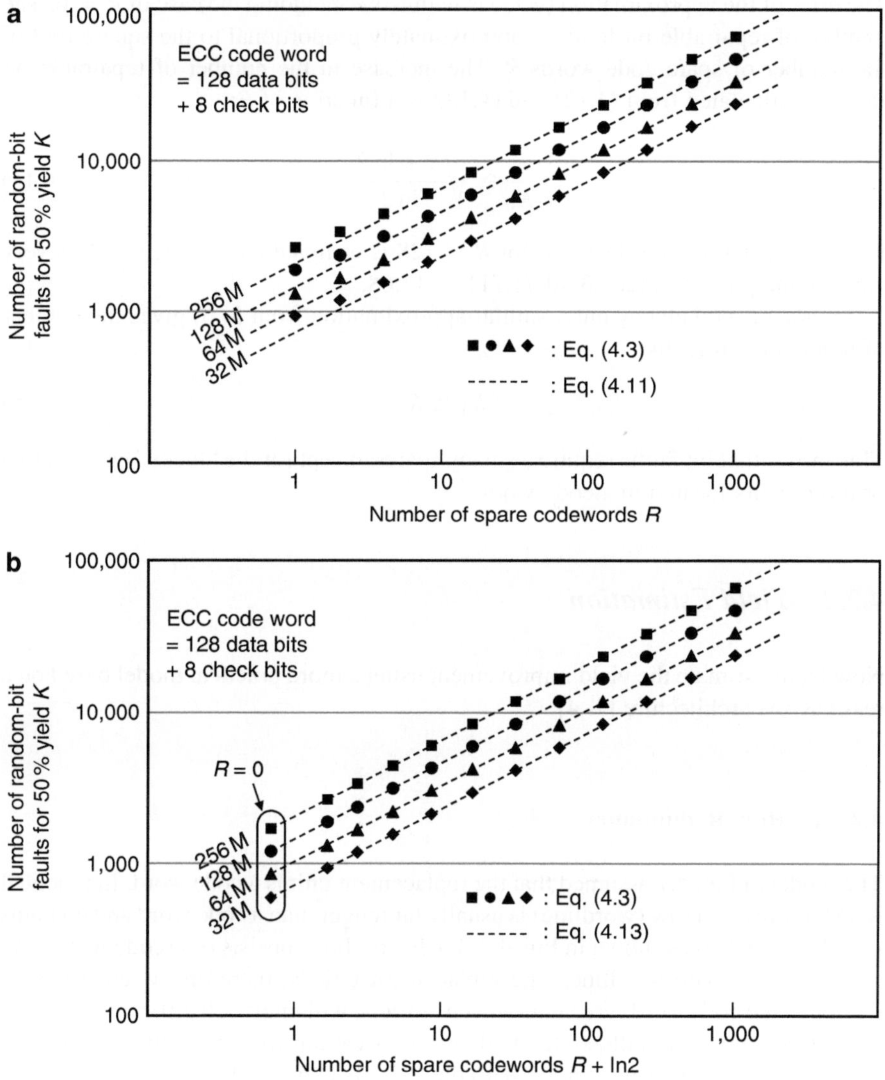

Fig. 4.3 Number of repairable random-bit faults using both redundancy and ECC: (**a**) approximation by (4.11) and (**b**) approximation by (4.13)

as described in Sect. 3.6. Comparing (4.11) and (4.12), we can find that the substitution of $R = \ln 2$ into (4.11) results in (4.12). Then, shifting the horizontal axis of Fig. 4.3a by ln 2, we obtain Fig. 4.3b, which is expressed as

$$K \cong \sqrt{\frac{2(R + \ln 2)Wn}{n - 1}}. \tag{4.13}$$

The error of the approximation (4.13) is within 3% including the case of $R = 0$. The number of repairable bit faults is approximately proportional to the square root of the number of spare code words R. The increase in the number of repairable bit faults is calculated from (4.12) and (4.13) as a function of R:

$$\frac{K}{K_0} \cong \sqrt{\frac{R + \ln 2}{\ln 2}}. \tag{4.14}$$

From this equation, the increase for $R = 128$ is estimated as 13.63, which is very close to the precise value 23,402/1,711 = 13.68.

In case of redundancy only, similar approximation from (4.6) gives the number of repairable bit faults:

$$K_1 \cong R. \tag{4.15}$$

This means that bit faults require as many spares except for the "lucky" cases of two or more faults located in a code word.

4.2.2 Yield Estimation

Now let us estimate the yield improvement using a more practical model based on a real DRAM architecture [1, 4].

4.2.2.1 Row Redundancy

The model in Fig. 4.1 assumed that the replacement unit is a code word. In practical RAM, however, a row (wordline) is usually far longer than a code word and contains plural code words as shown in Fig. 4.4. Each wordline consists of w code words and so does each spare wordline. The replacement unit is, therefore, w code words. Although the code words are "interwoven" to cope with multicell errors as described in Sect. 3.7, Fig. 4.4 is illustrated without interweaving for simplicity. In this case, the probability that a row is not repairable, p_{row}, is expressed as

$$p_{\text{row}} = 1 - (1 - p_{\text{cw}})^w, \tag{4.16}$$

where p_{cw} is the probability of a code word being nonrepairable expressed by (4.1). Using p_{row} instead of p_{cw} in (4.2) and (4.3), we obtain

$$P_i = \binom{W}{i} p_{\text{row}}^i (1 - p_{\text{row}})^{W-i}, \tag{4.17}$$

$$Y = \sum_{i=0}^{R} P_i = \sum_{i=0}^{R} \binom{W}{i} p_{\text{row}}^i (1 - p_{\text{row}})^{W-i}. \tag{4.18}$$

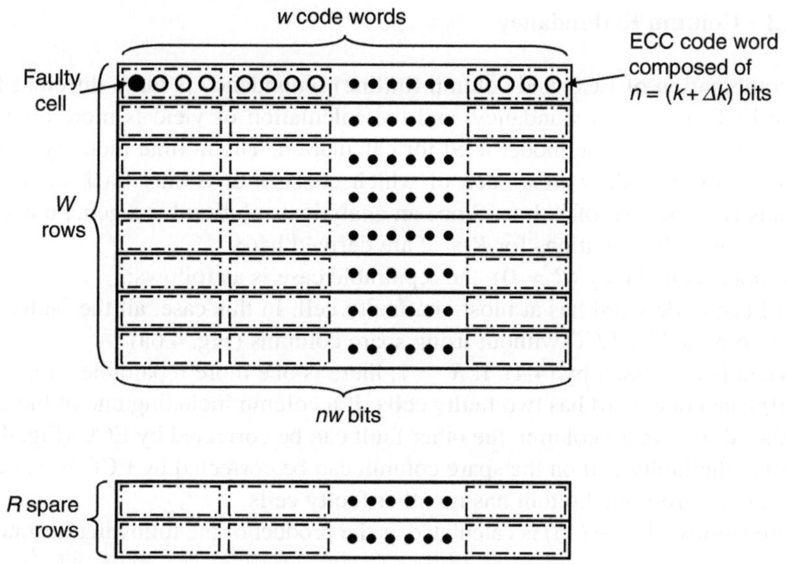

Fig. 4.4 Combination of row redundancy and ECC

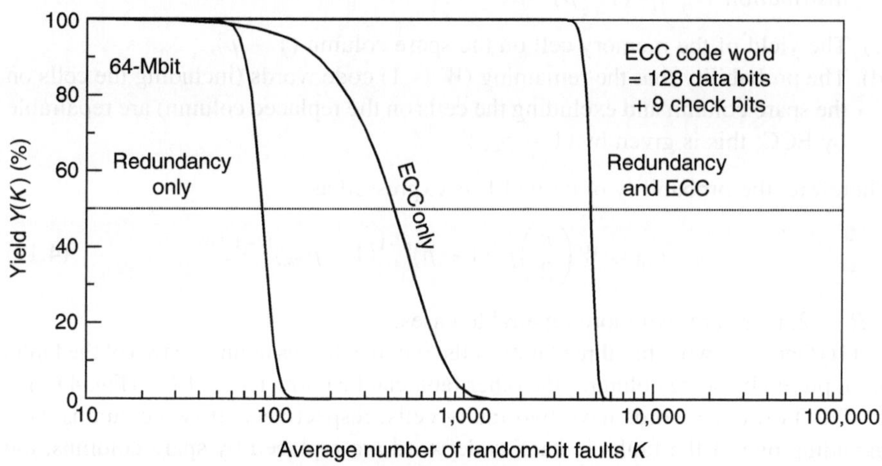

Fig. 4.5 Yield using row redundancy and ECC [1, 4]

In the ECC-only and redundancy-only cases, the yield is given by substituting p_{row} instead of p_{cw} into (4.4)–(4.7).

Figure 4.5 shows the calculation results of the yield of a 16-Mbit DRAM [1, 4], where $w = 8$, $W = 4{,}096$, and $R = 24$. Since the DRAM has four sets of the structure shown in Fig. 4.4, the yield is given by Y^4. A synergistic effect of about one decade is obtained. The calculation neglects faults on spare wordlines. If they are included, the yield will slightly decrease.

4.2.2.2 Column Redundancy

The combination of ECC and column (bitline) redundancy is quite different from that of ECC and row redundancy, and the calculation of yield is more complex. Figure 4.6 illustrates the model used for calculation. The normal memory array is composed of W code words, each of which consists of n bits. Each of R spare columns is composed of W bits. Since an analytical solution for a general R is too complex, only the equations for $R \leq 2$ are derived here.

Without redundancy ($R = 0$), the repairable case is as follows:

(0) Each code word has at most one faulty cell. In this case, all the faulty cells can be corrected by ECC without using spare columns (Fig. 4.6a).

The yield is expressed by (4.4). If $R = 1$, there is one more repairable case:

(1d) One code word has two faulty cells. If a column including one of the faults is replaced by a spare column, the other fault can be corrected by ECC (Fig. 4.6b). Note that the faulty cell on the spare column can be corrected by ECC because the third column from the bottom has no other faulty cells.

The probability of case (1d) is calculated as the product of the following factors [4]:

(a) The number of selection of a code word with two faults, W.
(b) The probability that the code word has two faults; this is given by the binomial distribution $\binom{n}{2} p^2(1-p)^{n-2}$.
(c) The yield of the memory cell on the spare column $(1 - p)$.
(d) The probability that the remaining $(W - 1)$ code words (including the cells on the spare column and excluding the cells on the replaced column) are repairable by ECC; this is given by $(1 - p_{\text{cw}})^{W-1}$.

Therefore, the probability of case (1d) is expressed as

$$Y_{1\text{d}} = W\binom{n}{2} p^2(1-p)^{n-1}(1-p_{\text{cw}})^{W-1}. \tag{4.19}$$

If $R = 2$, there are two more repairable cases:

(1t) One code word has three faulty cells. If two columns including two of the faults are replaced by spare columns, the other fault can be corrected by ECC (Fig. 4.6c).

(2d) Two code words have two faulty cells, respectively. If two columns each including one of the faults in each code word are replaced by spare columns, the other faults can be respectively corrected by ECC (Fig. 4.6d).

The probability of each case is calculated by similar calculation as case (1d) and is given by

$$Y_{1\text{t}} = W\binom{n}{3} p^3(1-p)^{n-1}(1-p_{\text{cw}})^{W-1}, \tag{4.20}$$

$$Y_{2\text{d}} = \binom{W}{2}\left\{\binom{n}{2} p^2(1-p)^{n-1}\right\}\left\{\binom{n-2}{2} p^2(1-p)^{n-1}\right\}(1-p_{\text{cw}})^{W-2}, \tag{4.21}$$

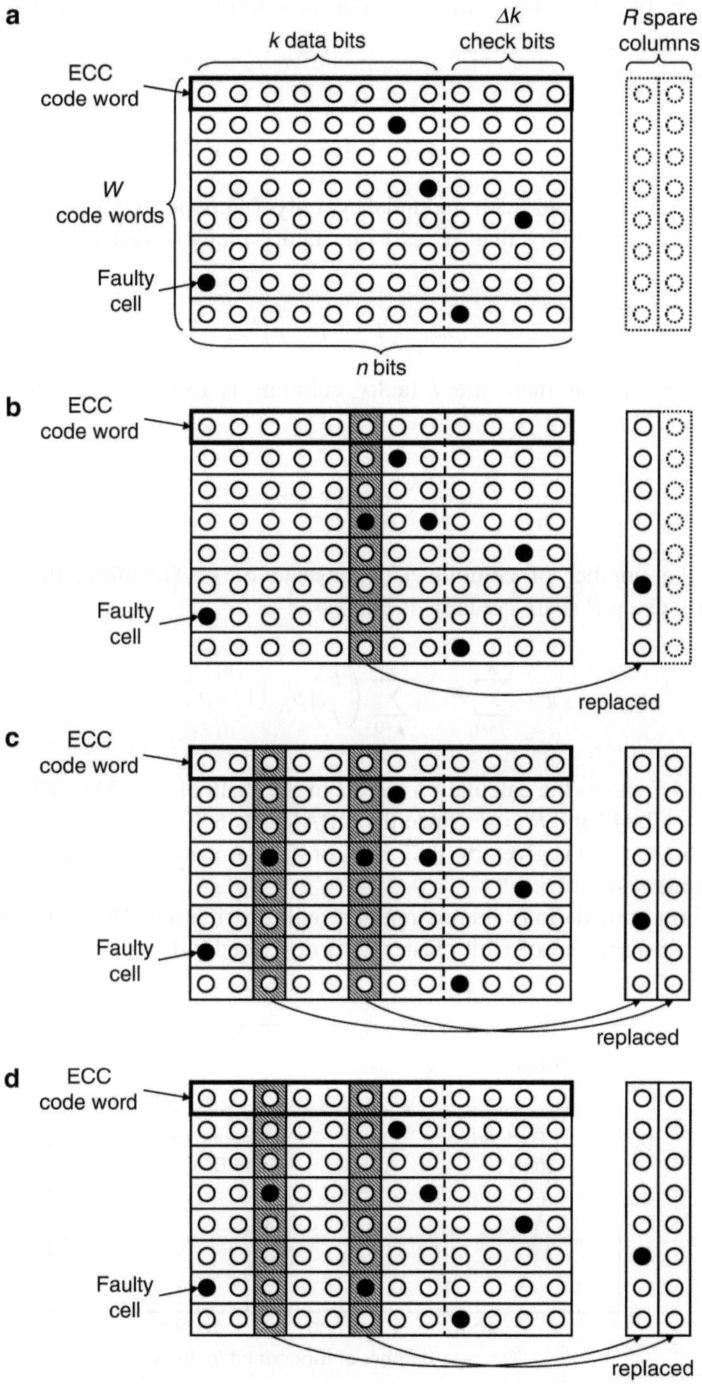

Fig. 4.6 Combination of column redundancy and ECC: (**a**) case (0) requires no spare columns; (**b**) case (1d) requires one spare column; (**c**) case (1t) requires two spare columns; (**d**) case (2d) requires two spare columns

respectively [4]. Therefore, the repairable probability for $R = 1$ and $R = 2$ are expressed as

$$Y_1 = Y_0 + Y_{1\mathrm{d}}, \tag{4.22}$$

$$Y_2 = Y_0 + Y_{1\mathrm{d}} + Y_{1\mathrm{t}} + Y_{2\mathrm{d}}, \tag{4.23}$$

respectively. The yield with redundancy only (without ECC) is calculated as follows. The probability that at least one faulty memory cell is in a column is given by

$$p_{\mathrm{col}} = 1 - (1-p)^w. \tag{4.24}$$

The probability that there are j faulty columns is expressed by the binomial distribution

$$P_j = \binom{k}{j} p_{\mathrm{col}}^j (1 - p_{\mathrm{col}})^{k-j}, \tag{4.25}$$

because the number of columns is k, instead of n. Therefore, the repairable probability using R spare columns is expressed as

$$Y_R = \sum_{j=0}^{R} P_j = \sum_{j=0}^{R} \binom{k}{j} p_{\mathrm{col}}^j (1 - p_{\mathrm{col}})^{k-j}. \tag{4.26}$$

Figure 4.7 shows the calculation results of the yield of a 16-Mbit DRAM [1, 4], where $W = 2{,}048$ and $R = 2$. Since the DRAM has 64 sets of the structure shown in Fig. 4.6, the yield is given by ${Y_2}^{64}$, ${Y_0}^{64}$, or ${Y_R}^{64}$. A synergistic effect of about six times is obtained.

The above calculations assume random fault distribution. The yield calculation that takes into account of fault clustering is described in [4].

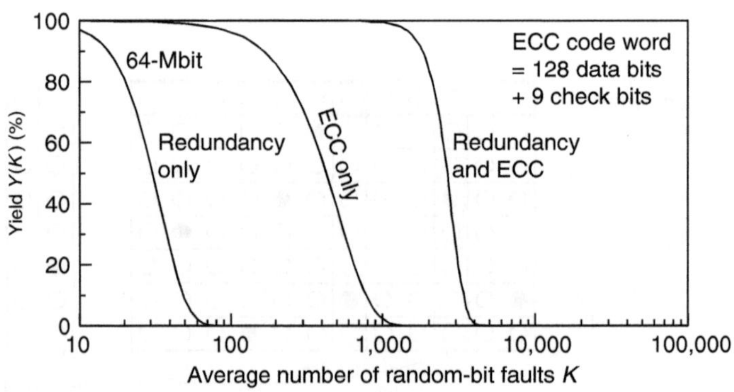

Fig. 4.7 Yield using column redundancy and ECC [1, 4]

4.2.2.3 Row and Column Redundancies

To the best of our knowledge, no analytical solutions have yet been found for the combination of ECC and both row and column redundancies. The yield, however, can be estimated using Monte Carlo simulation.

4.3 Application of Synergistic Effect

4.3.1 Threshold-Voltage Variations

In this section, variations of SRAM cells and DRAM sense amplifiers (SAs), which are the most promising applications of the synergistic effect, are described. The statistical variation of threshold voltage (V_t) is one of the most important problems for low-voltage memories using nanoscale devices. The V_t mismatch between pair transistors of an SRAM cell deteriorates the static noise margin of the cell [5, 6], while that of a DRAM SA reduces the effective signal voltage (Fig. 4.8). The mismatch is caused by the intrinsic V_t variation of the pair transistors due to random microscopic fluctuations of dopant atoms in the channel region [7]. Here, the standard deviation of the intrinsic V_t variation is given as

$$\sigma(V_t) = \frac{q}{C_{OX}}\sqrt{\frac{N_A D}{3LW}}, \tag{4.27}$$

where q is the electronic charge, C_{OX} is the gate-oxide capacitance per unit area, N_A is the impurity concentration, D is the depletion layer width under the gate, L is the

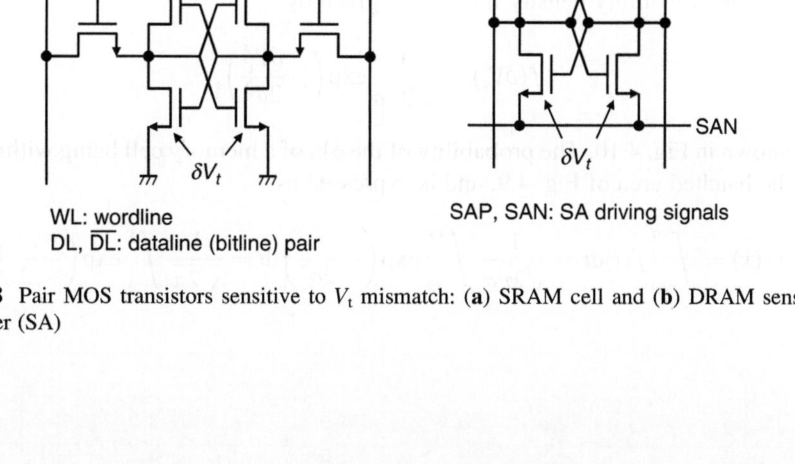

Fig. 4.8 Pair MOS transistors sensitive to V_t mismatch: (**a**) SRAM cell and (**b**) DRAM sense amplifier (SA)

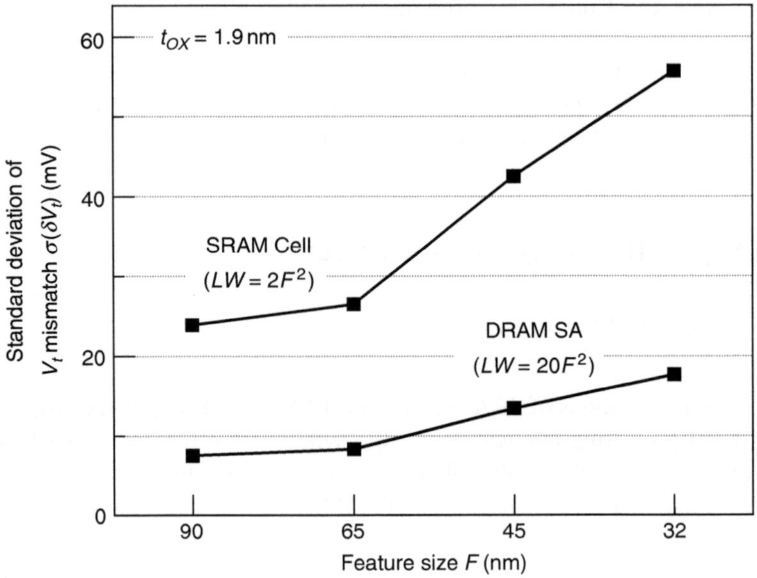

Fig. 4.9 Trend in standard deviation of V_t mismatch of SRAM cell transistors. Reproduced from [9] with permission; © 2010 IEEE

channel length, and W is the channel width [8]. It is obvious that the variation increases with device scaling because of a smaller LW and the ever-larger N_A for smaller short-channel effects. In practice, for smaller MOS transistors, N_A is not uniform in the channel region due to channel engineering. Therefore, empirical data are more useful to evaluate the variation. According to empirical data and device simulations for 1.9-nm t_{OX} transistors (t_{OX}: gate-oxide thickness), the standard deviation of the V_t mismatch, $\sigma(\delta V_t)$, is expected to increase with device scaling as shown in Fig. 4.9 [5, 9]. The maximum of the mismatch $|\delta V_t|_{max}$, however, depends not only on $\sigma(\delta V_t)$ but also on memory capacity M. The ratio between the maximum and the standard deviation $m = |\delta V_t|_{max}/\sigma(\delta V_t)$ is calculated as follows [10].

Let us start with SRAMs. The threshold-voltage mismatch δV_t is assumed to be expressed as a Gaussian distribution with a mean of zero and a standard deviation of σ. The probability density function is given by

$$f(\delta V_t) = \frac{1}{\sqrt{2\pi}\sigma}\exp\left(-\frac{\delta V_t^2}{2\sigma^2}\right), \tag{4.28}$$

as shown in Fig. 4.10. The probability of the δV_t of a memory cell being within $\pm x\sigma$ is the hatched area of Fig. 4.9, and is expressed as

$$Y_1(x) = \int_{-x\sigma}^{x\sigma} f(t)\mathrm{d}t = \frac{1}{\sqrt{2\pi}\sigma}\int_{-x\sigma}^{x\sigma}\exp\left(-\frac{t^2}{2\sigma^2}\right)\mathrm{d}t = \frac{1}{\sqrt{2\pi}}\int_{-x}^{x}\exp\left(-\frac{u^2}{2}\right)\mathrm{d}u. \tag{4.29}$$

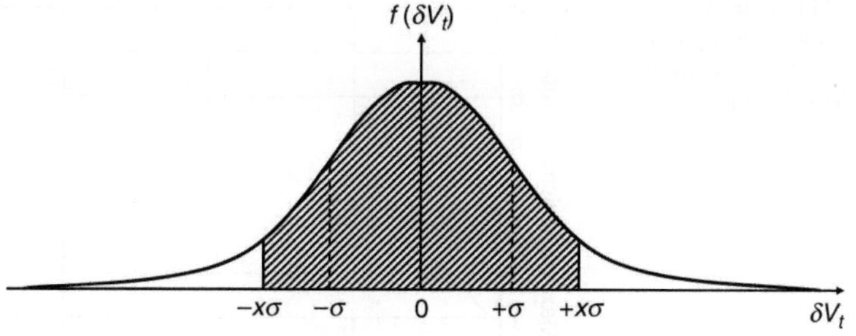

Fig. 4.10 Distribution of V_t mismatch δV_t

Fig. 4.11 Probability of the maximum V_t mismatch of an M-bit memory, $|\delta V_t|_{max}$, being within $x\sigma$ [10]

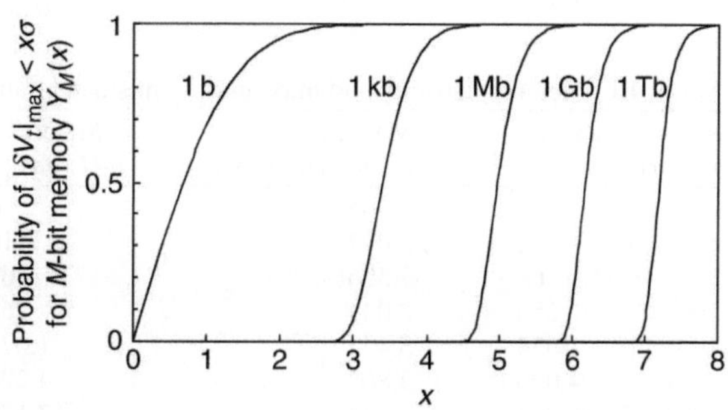

The probability of all the δV_ts of M memory cells being within $\pm x\sigma$, that is, the probability of $|\delta V_t|_{max} < x\sigma$, is given by

$$Y_M(x) = \{Y_1(x)\}^M. \tag{4.30}$$

Plotting (4.30) with M as a parameter results in Fig. 4.11. The ratio m between the expectation of $|\delta V_t|_{max}$ and the standard deviation is the area of left-hand side of this curve, and is expressed as

$$m = \int_0^\infty \{1 - Y_M(x)\}\mathrm{d}x = \int_0^\infty \left[1 - \left\{\frac{1}{\sqrt{2\pi}}\int_{-x}^{x} \exp\left(-\frac{u^2}{2}\right)\mathrm{d}u\right\}^M\right]\mathrm{d}x. \tag{4.31}$$

Figure 4.12 and Table 4.1 show the relationship between m and the memory capacity M. For example, $|\delta V_t|_{max}$ of a 256-Mbit memory is about six times the standard deviation.

4.3.2 Estimated Effect

The above estimation is for the case with neither redundancy nor ECC, as shown in the "$r = 0$" line in Fig. 4.13. However, repairing memory cells with excessive δV_t

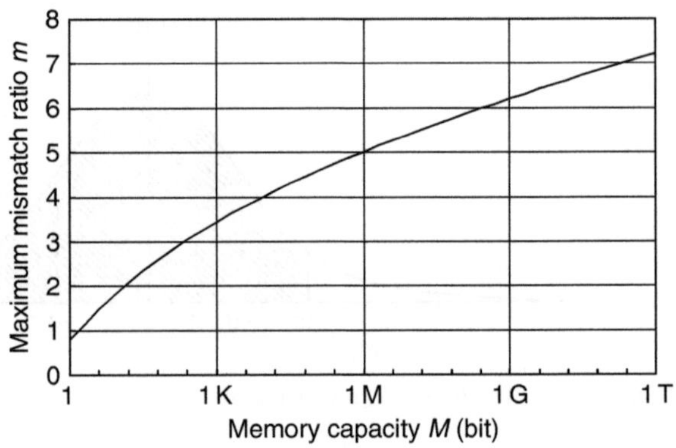

Fig. 4.12 Maximum V_t mismatch ratio m. Maximum V_t mismatch $|\delta V_t|_{max}$ is m times standard deviation [10]

Table 4.1 Memory capacity and maximum V_t mismatch ratio

Memory capacity M (bit)	Maximum mismatch ratio m	Memory capacity M (bit)	Maximum mismatch ratio m
4	1.465	4,194,304	5.269
16	2.077	16,777,216	5.518
64	2.596	67,108,864	5.757
256	3.044	268,435,456	5.987
1,024	3.442	1,073,741,824	6.209
4,096	3.802	4,294,967,296	6.424
16,384	4.134	17,179,869,184	6.631
65,536	4.443	68,719,476,736	6.833
262,144	4.733	274,877,906,944	7.029
1,048,576	5.008	1,099,511,627,776	7.220

by redundancy and/or ECC provides a smaller m and a smaller $|\delta V_t|_{max}$, which are calculated as follows. The probability of a memory cell with an excessive (larger than $m\sigma$) δV_t is calculated as

$$P(|\delta V_t|>m\sigma) = \int_{-\infty}^{-m\sigma} f(x)\mathrm{d}x + \int_{m\sigma}^{\infty} f(x)\mathrm{d}x = 1 - \frac{1}{\sqrt{2\pi}} \int_{-m}^{m} \exp\left(-\frac{u^2}{2}\right)\mathrm{d}u. \quad (4.32)$$

Therefore, $M{\cdot}P(|\delta V_t| > m\sigma)$ cells are expected to have excessive δV_ts. If all these cells are repaired by redundancy and/or ECC (i.e., repair ratio $r = P(|\delta V_t| > m\sigma)$), the maximum mismatch $|\delta V_t|_{max}$ is limited to $m\sigma$, because the remaining cells have smaller δV_ts. The ratio m is calculated from (4.32) so that $r = P(|\delta V_t| > m\sigma)$. The relationship between m and r is shown in Fig. 4.13. Unlike the no-repair case ($r = 0$), m is independent of M, because (4.32) does not contain M, while (4.31) does. For example, if 0.1% of memory cells are repaired, $|\delta V_t|_{max}$ is limited to 3.3σ.

The yield calculated using (4.1) and (4.3) is shown in Fig. 4.14a [9]. The number of required spare code words R to achieve a 50% yield is shown in Fig. 4.15a [9]. The figure also shows cases without ECC using (4.5) and (4.7), where all the

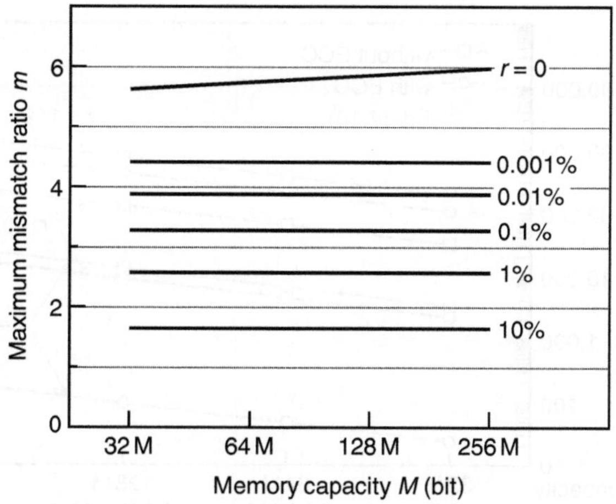

Fig. 4.13 Maximum V_t mismatch ratio m with repair rate of r. Reproduced from [9] with permission; © 2010 IEEE

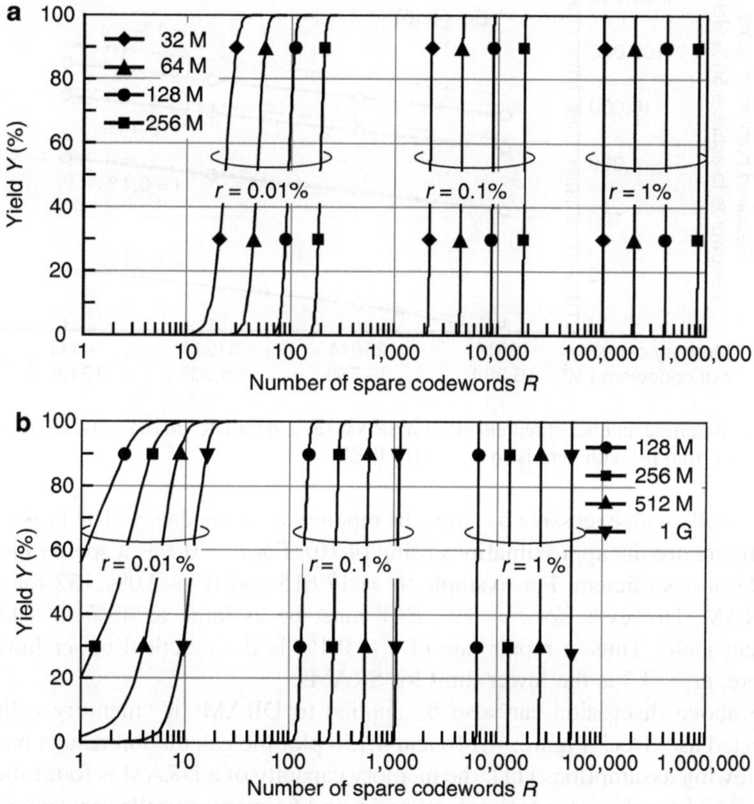

Fig. 4.14 Yield using both redundancy and ECC for SRAM (**a**) and DRAM (64 cells/SA) (**b**) [9, 11]. Reproduced from [9] with permission; © 2010 IEEE

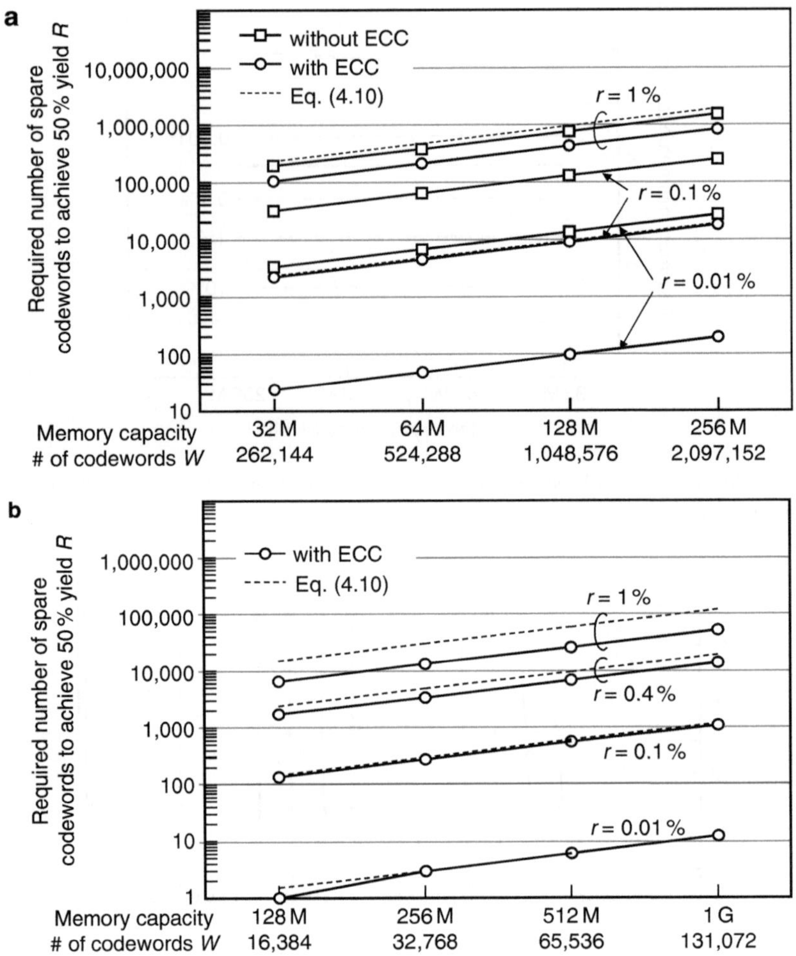

Fig. 4.15 Required number of redundancy for SRAM (**a**) and for DRAM (64 cells/SA) (**b**) [9, 11]. Reproduced from [9] with permission; © 2010 IEEE

memory cells with excessive δV_t must be repaired by redundancy. The broken lines in the figure are the approximations using (4.10). For $r = 0.1\%$, a spare ratio of $R/W = 0.9\%$ is sufficient. For example, $R = 17{,}615$ and $W = 2{,}097{,}152$ for a 256-Mb SRAM. However, for $r = 1\%$, R/W must be as large as 40% to achieve a sufficient yield. Thus, a repair rate of $r \approx 0.1\%$ is the practical upper limit and, therefore, $m \approx 3.3$ is the lower limit for SRAMs.

The above discussion can also be applied to DRAMs if "memory cells" are interpreted as "SAs." Figures 4.14b and 4.15b plot the calculation results based on the following assumptions [11]: the memory capacity of a DRAM is four times that of an SRAM for the same technology node, and 64 memory cells are connected to an SA. Consequently, the number of DRAM SAs is 1/16 the number of SRAM cells. Since r is inversely proportional to the square root of W for the same R (4.10), the DRAM's r can be four times as large as the SRAM's r. Thus, $r \approx 0.4\%$ is the upper limit ($m = 2.9$ is the lower limit) for DRAMs.

References

1. H. L. Kalter, C. H. Stapper, J. E. Barth Jr., J. DiLorenzo, C. E. Drake, J. A. Fifield, G. A. Kelley Jr., S. C. Lewis, W. B. van der Hoeven and J. A. Yankosky, "A 50-ns 16-Mb DRAM with a 10-ns data rate and on-chip ECC," IEEE J. Solid-State Circuits, vol. 25, pp. 1118–1128, Oct. 1990.
2. J. A. Fifield and C. H. Stapper, "High-speed on-chip ECC for synergistic fault-tolerant memory chips," IEEE J. Solid-State Circuits, vol. 26, pp. 1449–1452, Oct. 1991.
3. S.-H. Kim, W.-O. Lee, J.-H. Kim, S.-S. Lee, S.-Y. Hwang, C.-I. Kim, T.-W. Kwon, B.-S. Han, S.-K. Cho, D.-H. Kim, J.-K. Hong, M.-Y. Lee, S.-W. Yin, H.-G. Kim, J.-H. Ahn, Y.-T. Kim, Y.-H. Koh and J.-S. Kih, "A low power and high reliable 400Mbps mobile DDR SDRAM with on-chip distributed ECC," in Proc. ASSCC, Nov. 2007, pp. 34–37.
4. C. H. Stapper and H.-S. Lee, "Synergistic fault-tolerance for memory chips," IEEE Trans. Comput., vol. 41, pp. 1078–1087, Sep. 1992.
5. M. Yamaoka, K. Osada, R. Tsuchiya, M. Horiuchi, S. Kimura and T. Kawahara, "Low power SRAM menu for SOC application using yin-yang-feedback memory cell technology," in Symp. VLSI Circuits Dig. Tech. Papers, June 2004, pp. 288–291.
6. A. Agarwal, B. C. Paul and K. Roy, "Process variation in nano-scale memories: failure analysis and process tolerant architecture," in Proc. CICC, Oct. 2004, pp. 353–356.
7. M. J. M. Pelgrom, A. C. J. Duinmaijer and A. P. G. Welbers, "Matching properties of MOS transistors," IEEE J. Solid-State Circuits, vol. 24, pp. 1433–1440, Oct. 1989.
8. Y. Taur, D. A. Buchanan, W. Chen, D. J. Frank, K. E. Ismail, S.-H. Lo, G. A. Sai-Halasz, R. G. Viswanathan, H.-J. C. Wann, S. J. Wind and H.-S. Wong, "CMOS scaling into the nanometer regime," Proc. IEEE, vol. 85, pp. 486–504, Apr. 1997.
9. K. Itoh, M. Horiguchi and M. Yamaoka, "Low-voltage limitations of memory-rich nano-scale CMOS LSIs," in Proc. ESSCIRC, Sep. 2007, pp. 68–75.
10. K. Itoh, M. Horiguchi and H. Tanaka, *Ultra-low voltage nano-scale memories*, Springer, New York, 2007, Chapter 5.
11. K. Itoh and M. Horiguchi, "Low-voltage scaling limitations for nano-scale CMOS LSIs," Solid-State Electron., vol. 53, pp. 402–410, Apr. 2009.

Chapter 5
Reduction Techniques for Margin Errors of Nanoscale Memories

5.1 Introduction

The ever-increasing margin error, which results from reduced timing and voltage margins with device scaling under a given operating voltage V_{DD}, must be reduced sufficiently. Reduction in the minimum operating voltage V_{DD} (i.e., V_{min}) is the key to reduce the error, as described in Chap. 1. However, it has strictly been prevented by low-voltage scaling limitations [1–5] that are the major problems in the nanoscale era. The problems stem from two device parameters that are unscalable as long as conventional devices and circuits are used: the first is the high value of the lowest necessary threshold voltage V_t (i.e., V_{t0}) of MOSFETs needed to keep the subthreshold leakage low. Although many intensive attempts to reduce V_{t0} through reducing leakage have been made since the late 1980s [4–6], V_{t0} is still not low enough to reduce V_{DD} to the sub-1 V region. The second is the variation in V_t (i.e., ΔV_t), which becomes more prominent in the nanoscale era [1–5]. The ΔV_t caused by the intrinsic random dopant fluctuation (RDF) in the MOS channel is the major source of various ΔV_t components, as mentioned in Chap. 1. Unfortunately, it increases with device scaling, and intensifies many detrimental effects such as reduced timing and voltage margins of circuits, and increased soft error rates (SERs) in RAM cells and logic gates. Due to such inherent features of V_{t0} and ΔV_t, V_{min} stays at a high value and increases with device scaling. Hence, V_{DD} must be increased to maintain the margins in accordance with increasing V_{min}, which causes an increase in the power dissipation as well as degrades the device reliability due to increased stress voltage. Thus, V_{DD} is facing a 1-V wall in the 65-nm generation, and is expected to rapidly increase with further scaling of poly-Si bulk MOSFETs [3], as shown in Fig. 5.1. If increase in V_{DD} is not allowed, drastic reduction techniques for V_{DD} and thus V_{min} will be indispensable.

V_{min} and thus V_{DD} can be scaled down with device scaling while maintaining the margin to the same, if V_{t0} and ΔV_t are scaled down with innovative devices, circuits and subsystem designs, and the power supply integrity is ensured. This is because V_{DD} is the sum of V_{min}, ΔV, and ΔV_{ps}. Here, the ΔV is the sum of the

M. Horiguchi and K. Itoh, *Nanoscale Memory Repair*, Integrated Circuits and Systems 1,
DOI 10.1007/978-1-4419-7958-2_5, © Springer Science+Business Media, LLC 2011

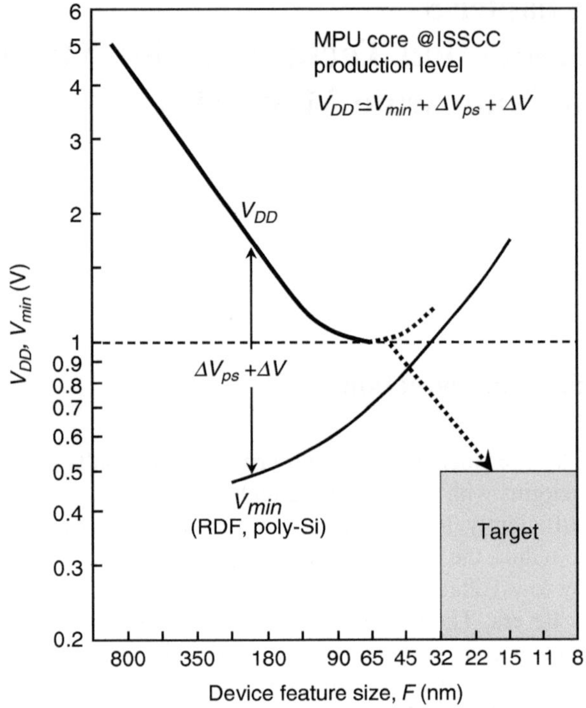

Fig. 5.1 Trends in V_{DD} and V_{min} of high-performance MPUs. Reproduced from [3] with permission; © 2010 IEEE

voltage needed to compensate for the extrinsic ΔV_t due to short-channel effects and line-edge roughness and of the voltage needed to meet the speed target. Thus, ΔV depends on the quality and maturity of the fabrication process, and the design target, which cannot be specified here. The ΔV_{ps} is the power supply droop and noise in the power supply lines and substrate. For example, the ever-higher resistance of interconnects in the nanoscale era [7–9] may affect power supply integrity by increasing ΔV_{ps}. As well, integrity depends on the chip packaging, such as 3D integration [10]. In any event, for the LSI industry in order to flourish and proliferate, the 1-V wall must be breached in the nanoscale era, while reducing V_{min} to maintain the timing and voltage margins to the same.

Concerns relating to adaptive circuits and relevant technologies to reduce V_{min} are addressed in this chapter. The focus is on memory-rich LSIs, as shown in Fig. 5.2. This is because such LSIs have usually driven the frontend of scaled devices development, and mixed signal and other types of LSIs sooner or later encounter similar problems. First, V_{min}, as a methodology to evaluate the low-voltage potential of MOSFETs, is proposed in terms of a tolerable speed variation, and the general features are described. Then, the V_{min}s of logic gates, SRAMs, and DRAMs are compared. After that, devices, circuits, and subsystems to widen the margins through reducing V_{min} are described.

a

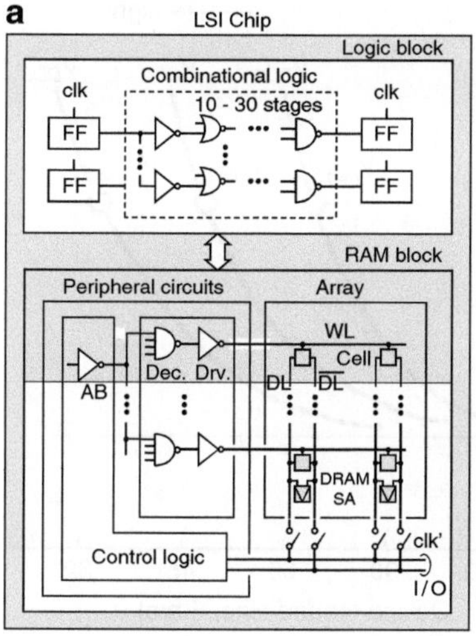

b

	Logic block	SRAM block		DRAM block		
		Periphery	Cells	Periphery	Sense amps	Cells
LW (av.)	4-12 F^2	4-12 F^2	1.5-2.5 F^2	4-12 F^2	10-20 F^2	1 F^2
V_t (av.)	0.2-0.4 V	0.2-0.4 V	0.2-0.7 V	0.2-0.4 V	0.2-0.4 V	0.7-1.3 V
t_{OX}	Thin	Thin	Usually thick	Usually thin	Thin	Thick
ΔV_t	Small	Small	Large	Small	Small	Large
Circuit count	Large	Small	Large	Small	Large	Large
Repair	No	No	Yes	No	Yes	Yes
Fan out	Small	Large*	–	Large*	–	–
Logical depth	Deep	Shallow	–	Shallow	–	–
Power off	Yes	Yes	No	Yes	No	No

Iterative-circuit sub-blocks

Fig. 5.2 (**a**) LSI composed of logic block and RAM block; (**b**) features of blocks. RAM block denotes SRAM block or DRAM block. Reproduced from [3] with permission; © 2010 IEEE

5.2 Definition of V_{min}

If a nanoscale MOSFET in a chip has an average V_t ($\cong V_{t0}$) with a maximum deviation (ΔV_{tmax}) in V_t from V_{t0}, the speed variation ($\Delta\tau$), that is, the ratio of the slowest speed at the highest V_t to the average speed at the average V_t ($\cong V_{t0}$) is approximately given (as discussed in Chap. 1) as [1]

$$\Delta\tau = \tau(V_{t0} + \Delta V_{tmax})/\tau(V_{t0}) = \{1 - \Delta V_{tmax}/(V_{DD} - V_{t0})\}^{-1.2}. \qquad (5.1)$$

It should be noted that $\Delta\tau$ stands for the timing margin under V_t-variations. Fortunately, for conventional MOSFETs, $\Delta\tau$ was negligible up till about the 130-nm-device generation because V_{DD} was much higher than V_{t0}, and ΔV_{tmax} was sufficiently small. In the nanoscale era below the 130-nm-device generation, however, $\Delta\tau$ has rapidly increased with device scaling due to the ever-increasing ΔV_{tmax}, as shown in Fig. 5.3. To offset the increase, V_{DD} must be increased, but this results in a continually increasing V_{DD} with device scaling. If V_{DD} is reduced under such circumstances, the increase in $\Delta\tau$ becomes catastrophic, as seen in (5.1). In practice, the increase in $\Delta\tau$ must be within a tolerable value ($\Delta\tau_0$) for reliable operation with a wide enough timing margin. If the minimum operating voltage (V_{min}) is defined as the V_{DD} necessary for achieving a tolerable $\Delta\tau_0$, V_{min} increases with device scaling, as seen in Fig. 5.3. V_{min} is obtained by solving (5.1) for V_{DD} as

$$V_{min} = V_{t0} + (1+\gamma)\Delta V_{tmax}, \quad \gamma = 1/({\Delta\tau_0}^{1/1.2} - 1), \quad \Delta V_{tmax} = m\sigma(V_t), \qquad (5.2)$$

$$\sigma(V_t) = A_{vt}(LW)^{-0.5}, \qquad \text{and} \quad A_{vt} \propto t_{ox}. \qquad (5.3)$$

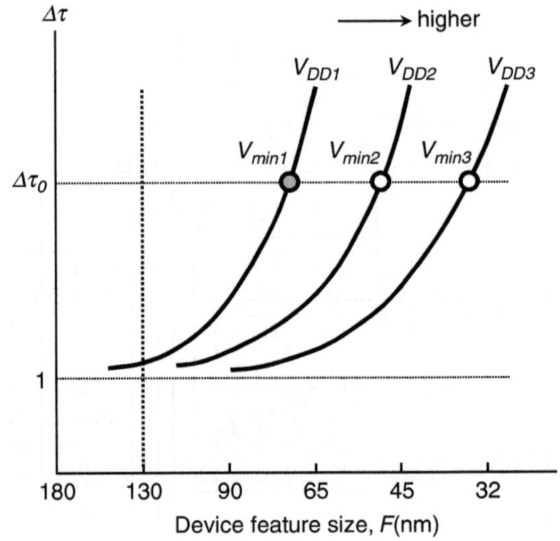

Fig. 5.3 Speed variation $\Delta\tau$ vs. device feature size, F [1, 11]. Reproduced from [11] with permission; © 2010 IEICE

For a conventional bulk MOSFET,

$$\sigma(V_t) = B_{vt}[t_{ox}(V_{t0} - V_{FB} - \Phi_S)/LW]^{0.5} \propto t_{ox}N_A{}^{0.25}(LW)^{-0.5}, \qquad (5.4)$$

where m depends on the circuit count in the block, $\sigma(V_t)$ is the standard deviation of V_t distribution, A_{vt} and B_{vt} are the Pelgrom and Takeuchi constants [12, 13], respectively, t_{ox} is the inversion electrical gate-oxide thickness, V_{FB} is the flat-band voltage, Φ_S is the surface potential, N_A is the impurity concentration of the channel, and LW is the MOSFET size. The $\Delta\tau_0$ can take two values, $\Delta\tau_0(+)$ and $\Delta\tau_0(-)$, corresponding to plus and minus values of ΔV_t. Here, $\Delta\tau_0(+)$ is used after this, simply expressed as $\Delta\tau_0$.

5.3 Reduction of V_{min} for Wider Margins

5.3.1 General Features of V_{min}

5.3.1.1 MOSFETs Governing V_{min}

The V_{min} of a chip is equal to the highest of V_{min}s of the three blocks (logic, SRAM, and DRAM) in the chip. The V_{min} of each block is governed by the circuit having the highest V_{min} in the block. Furthermore, the V_{min} of each circuit is governed by the MOSFET having the highest V_{min} in the circuit. Therefore, the V_{min} of each block is eventually determined by the MOSFET having the highest V_{min} in the block. Here, the MOSFET must be in a major core circuit impacting on power dissipation and speed of the block that are our major concerns in this chapter. Note that the smaller the MOSFET, the higher its V_{min} with a larger $\sigma(V_t)$. If a specific MOSFET is used more often in the block, causing an iterative circuit block, the V_{min} of the MOSFET is statistically higher with a larger ΔV_{tmax}. Furthermore, the larger the C_L/W (C_L: the load capacitance), the higher the V_{min} with a larger γ. This is because $\Delta\tau_0$ must be smaller as the C_L/W is large, so it becomes less influential in the block speed. For RAMs, the V_{min} is also influenced by operation modes, that is, the nondestructive read-out (NDRO) and thus ratio operations for SRAMs, and the destructive read-out (DRO) and refresh operations for DRAMs. Taking these general features into account, the MOSFET can be specified as M_1 in each circuit in Fig. 5.4. Note that the DRAM sense amplifier (SA) operates more simply than the SRAM cell for lack of any transfer MOSFET despite the same cross-coupled circuit configuration. The details are in what follows.

For the logic block, the statistical expression for ΔV_{tmax} in (5.2) has some ambiguity, unlike RAM blocks. Each gate does not work independently and randomly, and some gates form logical configurations with considerable logical depth and small fan out (Fig. 5.2b), enabling the $\sigma(V_t)$ to be reduced due to the averaging effect of random variations. The V_{t0} differs for some gates. For example, the well-known dual-V_{t0} logic block (BL) combines a low-V_{t0} MOSFET for the critical path and a high-V_{t0} MOSFET for the noncritical paths. The critical path tends to reduce the $\sigma(V_t)$ due to the low V_{t0} and large MOSFETs necessary to attain high speed. In addition, the small total MOS width of the path (typically about 10% of the total for the whole logic block) effectively reduces the m so that the noncritical paths inevitably determine the V_{min} of the whole block. Furthermore, the actual MOS size is different, ranging from 4 to $12F^2$ (F is feature size). To validate the equation even for such a logic block, however, it is assumed that the BL consists of many identical CMOS inverters, in which N/P MOSFETs have the same V_{t0} and size (i.e., $LW = 8F^2$ on average). The V_{min} of the BL calculated under these assumptions and using (5.2) may be higher than the actual V_{min}, including at least the averaging effect. This is also the case for peripheral logic circuits in RAM blocks because the circuit configurations are almost the same as those of the logic block. For array-relevant circuits in RAM blocks, however, the expression for

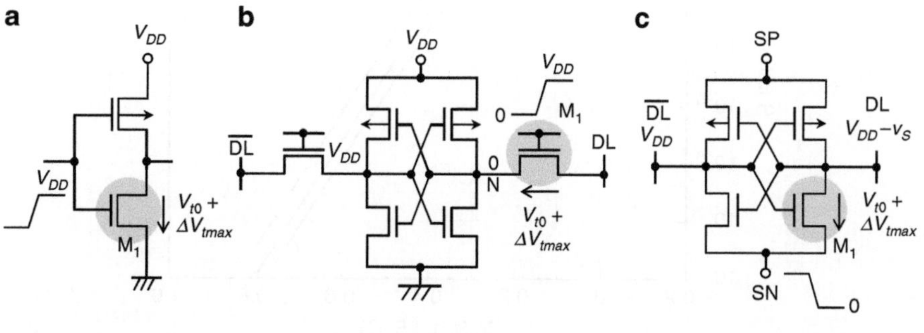

Fig. 5.4 (**a**) Inverter, (**b**) 6-T SRAM cell, and (**c**) DRAM sense amplifier. Reproduced from [3] with permission; © 2010 IEEE

ΔV_{tmax} is valid since each of the circuits comprises MOSFETs with the same V_{t0} and the same size, and operates independently and randomly.

For SRAMs using the six-transistor (6-T) cell, the V_{min} is equal to the highest of the three V_{min} values determined by cell stabilities at write and read, and tolerable speed variation at read. The V_{min} for write stability can be reduced sufficiently by power control of PMOSFET loads [14, 15] for a wider write margin, as explained later. The V_{min} for read stability can also be lowered by reducing the wordline voltage from V_{DD} [16], as is mentioned later. Hence, the V_{min} of SRAMs is determined by the speed variation of the transfer MOSFET. Unfortunately, the MOSFET always involves a slow and wide speed variation. The drawbacks come from the smallest MOSFET and the source voltage raised from ground (V_{SS}) level, caused by a ratio operation of the transfer and driver MOSFETs, and the largest V_{t0} variation due to the largest MOSFET count. The V_{min} can be calculated with (5.2) on the assumption that the source stays at 0 V during read operations, although this assumption makes the V_{min} lower than the actual V_{min} taking the raised source voltage into account. Here, the size of the transfer MOSFET is assumed to be $1.5F^2$. The V_{t0} is also assumed to be the same as those of cross-coupled MOSFETs shown in Fig. 5.5, since their leakage currents must be comparable in conventional designs.

For DRAMs, the DRO calls for restoring of the cell [4, 5] by utilizing the amplified signal by SA. It takes a long time because a small read signal must be amplified to a full V_{DD} on the heavily capacitive dataline, requiring a small

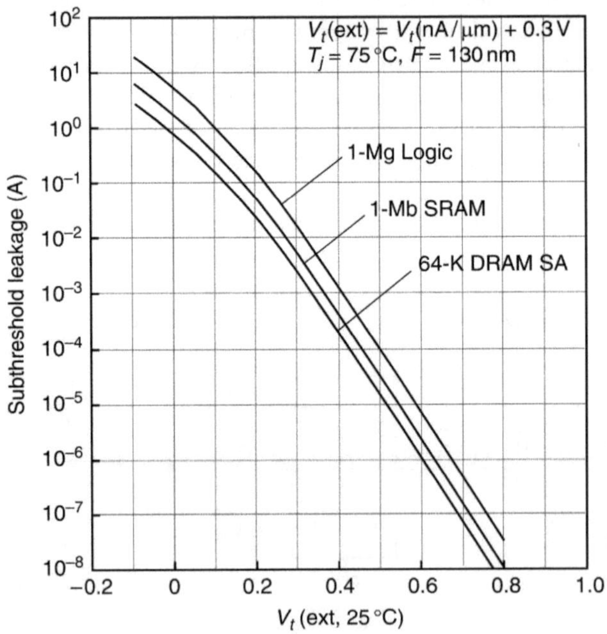

Fig. 5.5 Leakage vs. V_{t0} for various blocks. Reproduced from [3] with permission; © 2010 IEEE

$\Delta\tau_0$ and thus high $V_{\min}$ for confining the array speed within a tolerable value. Moreover, the refresh operation calls for simultaneous restoring of many cells along the selected wordline. This involves charging and discharging of many heavily capacitive datalines and operations of many SAs at a high voltage, causing high power. If the full-V_{DD} sensing (i.e., full-V_{DD} dataline precharging) is used, and activation of cross-coupled NMOSFETs in an SA precedes that of cross-coupled PMOSFETs [1, 4], (5.2) is applicable to M_1. However, the sensing imposes the use of the twin cell (two-transistor two-capacitor cell, i.e., 2T2C cell) that doubles the cell size of the conventional 1T1C, since generation of a stable and reliable reference voltage for signal discrimination [4] is difficult with the 1T1C cell, as discussed later. Note that other circuits are not core circuits that we have to pay attention to. For example, DRAM cells adopt the well-known word-bootstrapping to perform a full-V_{DD} write of the cell [4, 5], in which the wordline voltage is higher than the sum of the highest dataline voltage and the V_t of the cell transfer MOSFET. Therefore, to be exact, the word driver can have the highest $V_{\min}$ in the block. However, the driver quickly drives the wordline with a large MOSFET, and less contributes to power dissipation of the block because only one word diver is activated, unlike SAs. Moreover, in the past, DRAMs have solved the high-voltage problem by using high-voltage tolerant word drivers [4]. Furthermore, although the transfer MOSFET has the largest $\Delta V_{t\max}$ in the block due to the largest size/count, it never dominates the block speed. In fact, the developing speed of cell signal on the dataline is quick and insensitive to the V_t-variation thanks to a small voltage swing needed on the dataline and the word-bootstrapping. In any event, the full-V_{DD} sensing is assumed in the following, and the size of the NMOSFET is also assumed to be $15F^2$.

5.3.1.2 Lowest Necessary V_t (V_{t0})

The subthreshold current (leakage) of each off-MOSFET exponentially increases at the rate of about one-order increment per V_t-decrement of 100 mV because it is proportional to $W10^{-V_t/s}$ (W: channel width, S: subthreshold swing, $S \cong 100$ mV/decade) [5]. The lowest necessary V_t, V_{t0}, for the above-described MOSFETs depends on subthreshold-leakage specifications of the block or chip. Figure 5.5 plots the leakage vs. V_t and was prepared using previously reported data on cross-coupled MOSFETs in 130-nm SRAM cells [17]. Note that V_t is an extrapolated value that is familiar to circuit designers [5]. It is the sum of constant-current V_t (nA/μm) that is familiar to device designers and 0.3 V [5]. Moreover, the total MOS size, which governs the total subthreshold leakage, is assumed to be $16 \times 10^6F^2$ for 1-Mgate logic if a gate generates leakage from two MOSFETs with an average size of $8F^2$; $3.5 \times 10^6F^2$ for 1-Mb SRAM if a 6-T cell generates leakage from the two MOSFETs (total $LW = 3.5F^2$) of four cross-coupled MOSFETs in the cell; and $2 \times 10^6F^2$ for 64-k DRAM SAs, which contribute to leakage in the active standby mode (STB) if each SA generates leakage from two MOSFETs with an average LW of $15F^2$ for each. Obviously, the V_{t0} depends on the leakage specification. If the tolerated leakage is about 1–100 mA for a 1-Mgate logic block, 0.5–70 mA for a

1-Mb SRAM, and 0.15–20 mA for 64-k DRAM SAs, the V_{t0} is between 0.2 V (for high-speed designs) and 0.4 V (for low-power (LP) designs). Note that the leakage of the chip increases as logic gate and memory integration in the chip further increases. In this case, the V_{t0} must be increased so as to meet the specification. If the leakage is reduced, the V_{t0} can be reduced. Reduction in V_{t0}, however, requires the development of innovative low-leakage circuits and subsystems [1–5] as discussed later.

5.3.1.3 Parameter γ

This parameter strongly depends on the tolerable speed variation, $\Delta\tau_0$. In general, the BL needs a small $\Delta\tau_0$ (i.e., large γ) because the timing control must be quickly and stringently managed so as to meet the targeted speed from one flip flop (FF) to the other at every combinational logic stage (Fig. 5.2a). The speed is usually one-clock latency when measured in terms of the necessary number of clocks. In contrast, for RAM blocks, such a quick and stringent timing control is extremely difficult because of a large physical memory array, which inherently contains large delay components throughout the array. This difficulty occurs in the SRAM cell and the DRAM SA, each of which dominates the block speed with a large C_L/W. For example, a small transfer MOSFET in an SRAM cell must discharge a heavily capacitive data (bit) line, which takes a long time. Unfortunately, the discharging time varies greatly due to a wide variation in the V_t of the MOSFET and the ratio operation. The discharging signal must be aligned with a column clock (clk′ in Fig. 5.2a), waiting for the signal from the slowest cell so that the signal transferred to I/O is discriminated correctly. Such an operation unavoidably tolerates a two-clock latency, as typically seen in actual designs, as a result of giving up a one-clock latency that requires an extremely high V_{min} to offset the speed variation. This is also the case for DRAM SAs. Therefore, $\gamma = 3.09$ and $\Delta\tau_0 = 1.4$ for the BL and $\gamma = 2.09$ and $\Delta\tau_0 = 1.6$ for the SRAM and DRAM blocks are used here, with practical designs taken into account.

5.3.1.4 Maximum Deviation, ΔV_{tmax}

The number m ranges from 4.9 to 6.0 for 0.6- to 320-Mgate logic blocks from 5.2 to 6.3 for 4-Mb to 2-Gb SRAMs and from 4.8 to 5.9 for the 16-Mb to 8-Gb DRAMs connecting 64 cells to an SA [4]. It also depends on the repairable percentage, r, for RAMs. For the upper limit of r (i.e., 0.1% for SRAMs and 0.4% for DRAMs), attained by a combination of error correcting code (ECC) and redundancy, m is reduced to about 3.29 for SRAMs and to about 2.88 for DRAMs [1, 2], as described in Chap. 4. Note that, for a conventional bulk MOSFET, $\sigma(V_t)$ also depends on V_{t0}, as seen from (5.4). For $V_{FB} = -0.9$ V and $\Phi_S = 0.8$ V, $\sigma(V_t)$ is reduced to 0.45 of $\sigma(V_t = 0.4\text{ V})$, when V_{t0} is reduced from 0.4 to 0 V, as seen in Fig. 5.17. Furthermore, $\sigma(V_t)$ depends on A_{vt} and F^2, as expected from (5.3).

5.3.2 Comparison of V_{min} for Logic Block, SRAMs, and DRAMs

Figure 5.6 compares the $V_{\min}$ calculated for the BL and repaired RAMs for $A_{vt} = 4.2$ mV·μm [3]. The $V_{\min}$s of the logic and SRAM blocks are almost the same but still high, reaching an intolerable level of about 1.5 V in the 32-nm generation. Obviously, the $V_{\min}$ of DRAMs is the lowest due to the smallest $\sigma(V_t)$ and fewer SAs. The prime concern is thus the SRAM because its $V_{\min}$ is actually the highest when repair techniques are not used and the raised cell-node voltage is taken into consideration. In any event, the ever-increasing $V_{\min}$ with device scaling must be reduced. Thus, more advanced MOS structures, circuits, and even subsystems for logic, SRAM, and DRAM blocks must be developed. Otherwise, their timing and voltage margins will be fatal. The details are described in the following.

5.4 Advanced MOSFETs for Wider Margins

One of the most effective ways to reduce $V_{\min}$ is to use small A_{vt} and thus small $\sigma(V_t)$ MOSFETs. In the past, A_{vt} has been reduced as a result of using advance MOSFETs, as shown in Fig. 5.7. For 130-nm poly-Si gate bulk NMOSFETs [18, 19], A_{vt} was about 4.2 mV·μm when V_{t0} and t_{ox} were 0.30–0.45 V and 2.1–2.4 nm, respectively. The most advanced planar MOSFETs in the 45-nm generation have a low A_{vt} of 1.5–2.5 mV·μm [20–22] with high-*k* metal-gate materials for a thinner t_{ox} and/or a fully depleted (FD) silicon-on-insulator (FD-SOI) structure for a smaller N_A. Note

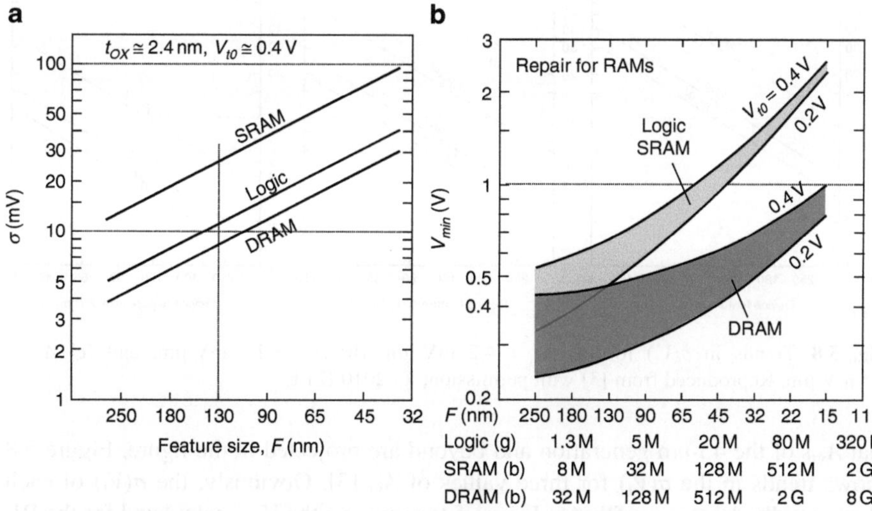

Fig. 5.6 Comparisons of (**a**) $\sigma(V_t)$ and (**b**) $V_{\min}$ of three circuit blocks [3]. $A_{vt} = 4.2$ mV·μm, $LW = 8F^2$ (logic), $1.5F^2$ (SRAM), and $15F^2$ (DRAM)

Fig. 5.7 Trends in t_{ox} and A_{vt}. Reproduced from [3] with permission; © 2010 IEEE

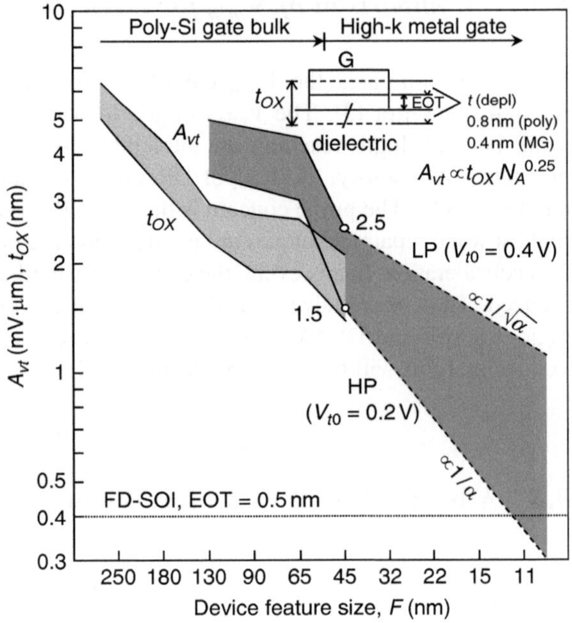

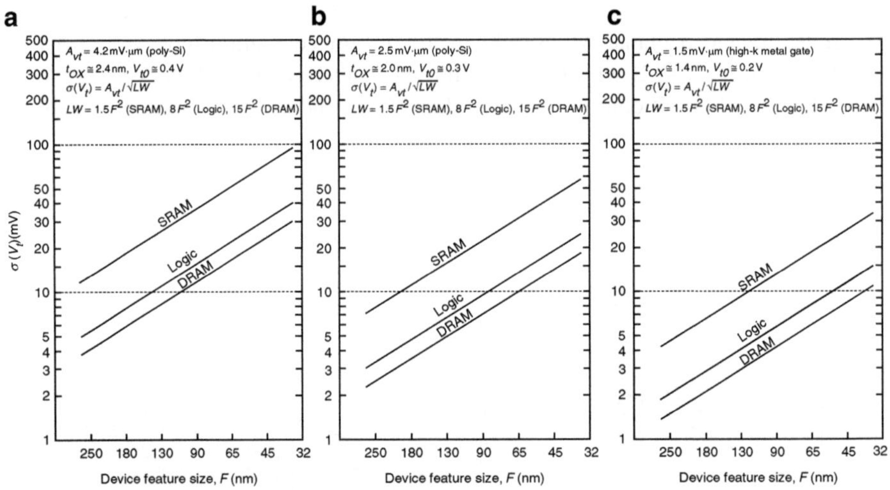

Fig. 5.8 Trends in $\sigma(V_t)$ for (**a**) $A_{vt} = 4.2$ mV·μm, (**b**) $A_{vt} = 2.5$ mV·μm, and (**c**) $A_{vt} = 1.5$ mV·μm. Reproduced from [3] with permission; © 2010 IEEE

that A_{vt}s of the 45-nm generation and beyond are projected in the figure. Figure 5.8 shows trends in the $\sigma(V_t)$ for three values of A_{vt} [3]. Obviously, the $\sigma(V_t)$ of each block rapidly decreases with A_{vt}. Figure 5.9 compares the V_{min} calculated for the BL and repaired RAMs for three values of A_{vt} [3], showing the strong dependence of V_{min} on A_{vt}. For $A_{vt} = 4.2$ mV·μm, the V_{min}s of the logic and SRAM blocks are almost

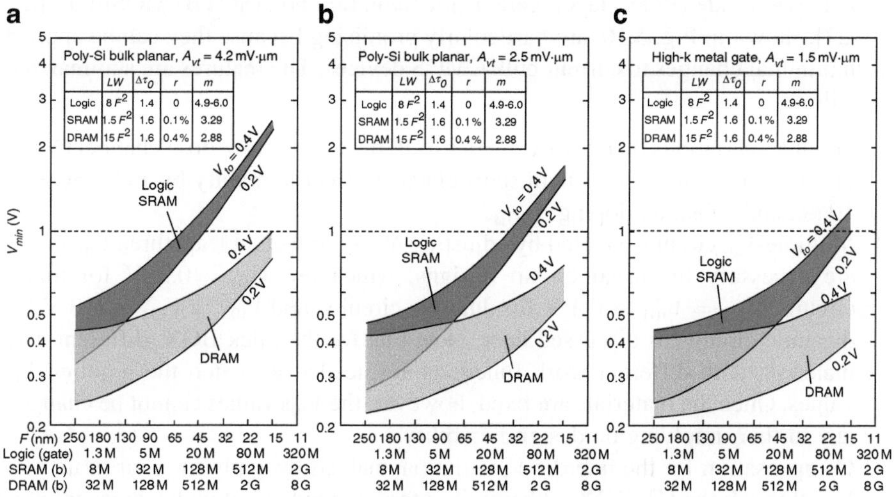

Fig. 5.9 $V_{\min}$s for the logic block and repaired RAMs for various MOSFETs having (**a**) $A_{vt} = 4.2$ mV·µm, (**b**) $A_{vt} = 2.5$ mV·µm, and (**c**) $A_{vt} = 1.5$ mV·µm. Reproduced from [3] with permission; © 2010 IEEE

the same but still high, reaching an intolerable level of about 1.5 V in the 32-nm generation. For $A_{vt} = 1.5$ mV·µm, however, they are reduced to less than 1 V even in the 22-nm generation.

FD-SOI structures, such as planar FD-SOI and fin-type MOSFETs (FinFETs), seem to be vital as future low-$V_{\min}$ MOSFETs due to their small A_{vt}s, although the fabrication processes have not been established yet, and the real problems thus remain unclarified. Their exceptional features are summarized as follows:

1. Small $\sigma(V_t)$ due to an ultra-low-dose channel.
2. Negligible p–n junction leakage due to no p–n junction at the source and drain. Long-refresh-time DRAMs and reliable on-chip-V_{BB} controls as discussed in Chap. 6 are thus enabled.
3. Small SERs due to ultrathin silicon.
4. Small parasitic capacitances due to its inherent structure surrounded by insulators.
5. No body effect enabling to speed up source-follow mode and reduce necessary wordline voltage of DRAM cells.
6. Built-in-large channel-width structure of FinFETs enabling high-speed high-density devices, such as DRAM cells and capacitor. The side wall process further enhances the features of FinFETs, as explained later.

5.4.1 *Planar FD-SOI MOSFETs*

Recently, considerable development effort has been directed toward planar FD-SOI devices as well as FinFETs [23]. Of these, FD-SOI MOSFETs with an ultrathin

(UT) buried oxide (BOX) layer, called silicon on thin box (SOTB) MOSFETs [22, 24–27], shown in Fig. 5.10, are particularly promising because they can be applied with minimal changes to current bulk CMOS devices. The features are summarized as follows.

1. Small V_t-variations thanks to a smaller RDF by an ultra-low-dose channel (N_{SOI}) of 10^{16} cm^{-3}, and an excellent short-channel-effect immunity by a UT-BOX of 10 nm and substrate doping N_{SUB}.
2. Multiple-V_{t0} circuits enabled by adjusting N_{SUB}. In fact, at least three V_{t0} values are necessary for usual circuit designs, which are $V_{t01} = 0.3$ V for major circuits, $V_{t02} = V_{t01} - 0.1$ V for dual-V_{t0} circuits, and $V_{t03} = V_{t01} - 0.3$ V for dynamic circuits as discussed later. Note that for the thick BOX, different gate materials with different work functions are needed to match the required V_{t0} values. Once the materials are fixed, however, the V_{t0}s values cannot be changed any further, implying no design flexibility.
3. Compensation for the interdie V_t variation enabled by applying a substrate bias V_{BB} through the UT-BOX. Otherwise, interdie speed variation becomes intolerable, as explained in Chap. 6.

Comparisons of A_{vt} between planar FD-SOIs and bulk MOSFETs are made in Fig. 5.11 [26]. For a given t_{ox}, the A_{vt} of FD-SOI is smaller due to the ultra-low-dose channel. Note that within-wafer variations are large even for FD-SOIs, and interdie V_t variations must thus be compensated for, as discussed in Chap. 6. Figure 5.12 compares V_{min}s of 4-Mb SRAMs using bulk and SOTB MOSFETs [28]. Figure 5.12a depicts experimental cumulative probability of the cells. For the bulk, V_{min} is about 0.8 V while 0.6 V for SOTB, showing a small difference of 0.2 V. This is because the channel length L was 90 nm for the bulk, but 50 nm for SOTB. Figure 5.12b shows the simulated probability for the bulk after calibration of L. Obviously, SOTB drastically reduces V_{min} from 1 to 0.6 V for a given channel length.

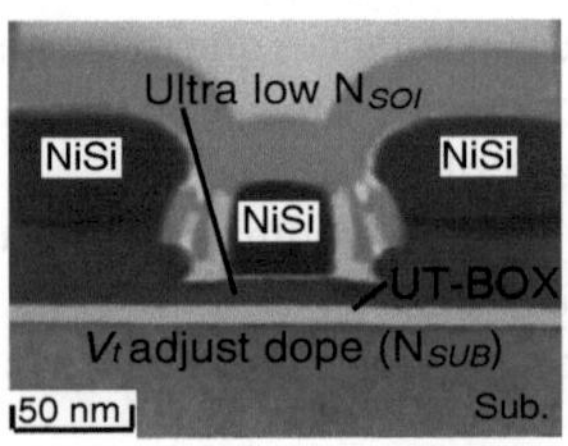

1. Small RDF due to ultra low-dose channel (N_{SOI})
2. Excellent SCE by ultra-thin (UT) BOX and N_{SUB}
3. Multiple V_t by adjusting substrate doping (N_{SUB})
4. Variation reduction and power optimization by V_{BB}
5. I/O bulk trs. can be easily integrated by removing BOX

Fig. 5.10 Schematic cross section of SOTB and its features. Reproduced from [26] with permission; © 2010 IEEE

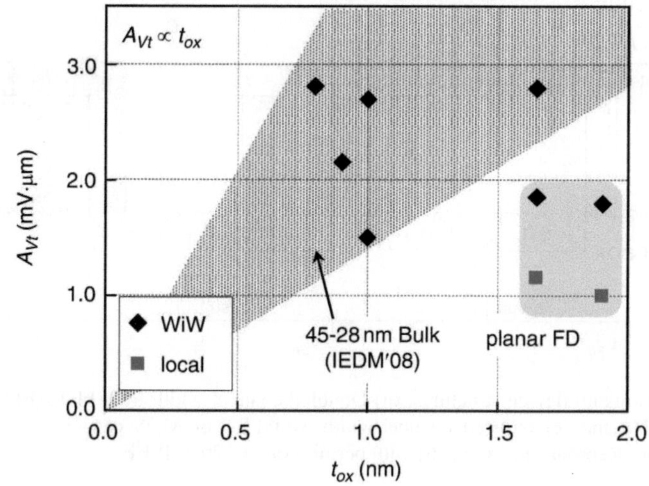

Fig. 5.11 Comparison of Pelgrom coefficient A_{vt} for NMOS. WiW within-wafer variation. Reproduced from [26] with permission; © 2010 IEEE

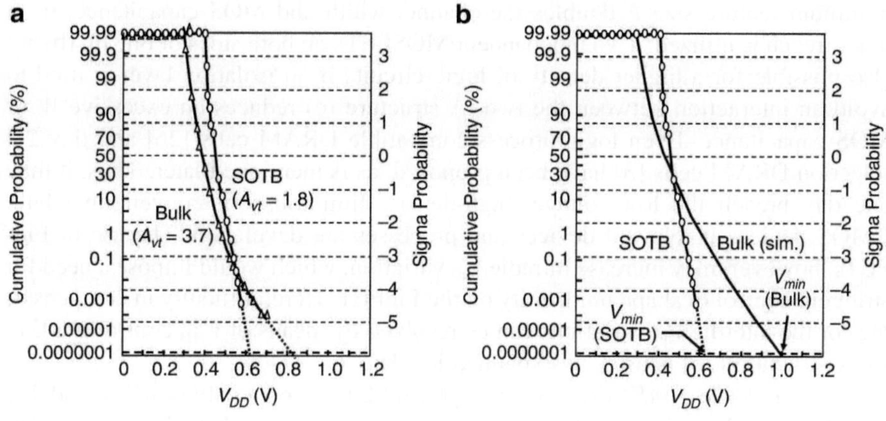

Fig. 5.12 Comparisons of V_{min}s of 4-Mb bulk and SOTB SRAMs: (**a**) before calibration (experiment) of gate length; $L = 90$ nm (bulk) and 50 nm (SOTB), and (**b**) after calibration of gate length; $L = 50$ nm (bulk and SOTB). Reproduced from [28] with permission; © 2010 IEEE

5.4.2 FinFETs

FinFETs feature an ultra-low-dose channel and a wide-channel-width built-in structure [11, 26]. Thus, it achieves not only a higher density and higher drive

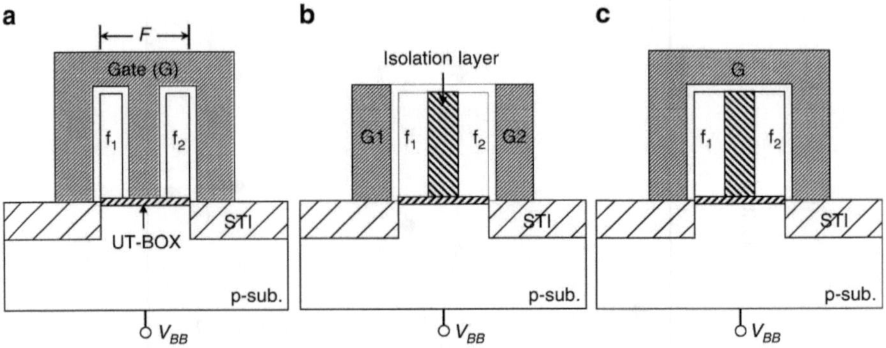

Fig. 5.13 Various fin-device structures. (**a**) Doubled channel-width MOSFET, (**b**) two independent MOSFETs, and (**c**) reduced channel-width MOSFET or MOS capacitance. *STI* shallow trench isolation. Reproduced from [26] with permission; © 2010 IEEE

current but also minimizes $\sigma(V_t)$ with minimized A_{vt} and maximized W. For example, if a side wall process [29] is used, further advanced MOSFETs and high-density capacitors [26] are achieved, as shown in Fig. 5.13. Structure (a) in the figure that accommodates two subfins (f1, f2) in one fin whose width is the minimum feature size F doubles the channel width and MOS capacitance, if an inner trench is utilized. Two independent MOSFETs on both sides of one fin (b) are also possible for a higher density of logic circuits if an isolation layer is used to avoid an interaction between the two. A structure (c) reduces an excessive W or MOS capacitance. Even logic-process-compatible DRAM cells [26] and tiny 2D selection DRAM cells [3] have been proposed, as is mentioned later. Thus, it may one day breach the low-voltage, high-density limitations of conventional bulk CMOS devices if relevant devices and processes are developed. The use of FinFETs, however, may increase intradie V_{t0} variation, which would impose a need for stringent control of shape uniformity on the FinFET. Here, difficulty in compensating for the interdie V_{t0} variations can be resolved by means of V_{BB} control if a UT-BOX structure [26] is used, as explained in Chap. 6.

$\sigma(V_t)$-Scalable FinFETs: The use of FinFETs even enables $\sigma(V_t)$-scalable MOSFETs to be achieved. If all feature sizes of a planar MOSFET are scaled down by a factor of $1/\alpha$ (α: device scaling factor > 1), as illustrated in Fig. 5.14, and the scaling factor of A_{vt} is reduced with $\alpha^{-0.5}$, as exemplified by LP designs in Fig. 5.7, $\sigma(V_t)$ and thus V_{min} increases by a factor of $\alpha^{0.5}$. Furthermore, $\sigma(V_t)$ and thus V_{min} remains constant even with A_{vt} scaling as large as α^{-1} (i.e., for high-performance (HP) designs in Fig. 5.7). However, the vertical structure provided by FinFETs [3, 11] yields a new scaling law for $\sigma(V_t)$, mitigating the requirement to A_{vt}. This is because this structure enables LW to be kept constant or even increased when the fin height (i.e., channel width W) is scaled up despite channel length L being scaled down. This can be done without sacrificing MOSFET density. This up-scaling is done in accordance with the degree of A_{vt} scaling, so $\sigma(V_t)$ and thus V_{min} are scaled down. For example, if A_{vt} is scaled down at $\alpha^{-0.5}$,

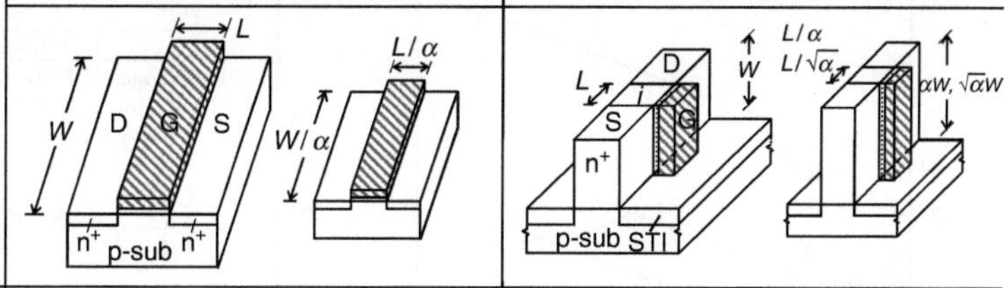

		Planar MOSFET	FinFET	
L		$1/\alpha$	$1/\alpha$	$1/\sqrt{\alpha}$
W		$1/\alpha$	α	$\sqrt{\alpha}$
W/L		1	α^2	α
LW		$1/\alpha^2$	1	1
A_{vt}		$1/\sqrt{\alpha}\,(1/\alpha)$	$1/\sqrt{\alpha}\,(1/\alpha)$	$1/\sqrt{\alpha}\,(1/\alpha)$
$\sigma(V_t)$		$\sqrt{\alpha}\,(1)$	$1/\sqrt{\alpha}\,(1/\alpha)$	$1/\sqrt{\alpha}\,(1/\alpha)$
V_{DD}		$\sqrt{\alpha}\,(1)$	$1/\sqrt{\alpha}\,(1/\alpha)$	$1/\sqrt{\alpha}\,(1/\alpha)$
I_{DS}		$\sim\alpha^{1.1}$	$\sim\alpha^{1.9}$	$\sim\alpha^{0.9}$
$\tau(MOS)$		$\sim\alpha^{-2.1}$	$\sim\alpha^{-1.9}$	$\sim\alpha^{-0.9}$
$P\ (= V_{DD}I_{DS})$		$\sim\alpha^{1.6}$	$\sim\alpha^{1.4}$	$\sim\alpha^{0.4}$
$P\tau(MOS)$		$\sim\alpha^{-0.5}$	$\sim\alpha^{-0.5}$	$\sim\alpha^{-0.5}$
W_{min}/L_{min}	$F = 45$ nm ($\alpha = 1$)	45/45 nm	45/45 nm	45/45 nm
	$F = 11$ nm ($\alpha = 4$)	11/11 nm (aspect ratio = 1)	180/11 nm (16)	90/23 nm (4)

$A_{vt} \propto t_{OX}N_A^{0.25}$, $\sigma(V_t) = A_{vt}/\sqrt{LW}$, $I_{DS} = \beta(V_{DD} - V_t)^{1.2}$ for constant N_A, $\tau(MOS) = V_{DD}C_G/I_{DS}$

Fig. 5.14 Comparisons of scaling between planar MOSFET and FinFET. Reproduced from [3] with permission; © 2010 IEEE

$\sigma(V_t)$ can also be scaled down by the same factor because LW is preserved as a result of the factor of α^{-1} or $\alpha^{-0.5}$ for L, and α or $\alpha^{0.5}$ for W. Such FinFETs enable high-speed operation not only due to the large drive current but also due to the shorter interconnects deriving from the vertical structures. However, the aspect ratio (W/L) of FinFETs increases with device scaling. For example, it is as large as 4–16 in the 11-nm generation, as shown in Fig. 5.14. Note that structures with such large aspect ratios might be possible to achieve, taking the history of DRAM development into account. In DRAMs, the aspect ratio of trench capacitors has increased from about 3 in the early 1980s to as much as 70 for modern 70-nm DRAMs [30, 31]. In addition, the ratio is almost halved for a given W if a sidewall process [29] is used. Even if the resultant W is still unnecessarily wide and thus increases power dissipation for a load dominated by the gate capacitance, the sidewall process further halves the W by splitting a MOSFET into two independent MOSFETs [26] (Fig. 5.13). For a load dominated by wiring capacitance, the large W due to FinFETs enables a high speed.

On the basis of the MOSFET scaling, we can predict the V_{min} for future blocks, assuming that the A_{vt} in the 45-nm generation, the A_{vt} scaling factor for further device scaling, and V_{t0} are 2.5 mV µm, $\alpha^{-0.5}$, and 0.4 V for LP designs, and 1.5 mV µm, α^{-1}, and 0.2 V for high-performance designs (Fig. 5.7). The constant LW in Fig. 5.14 is also assumed for FinFETs. Obviously, the use of FinFETs enables $\sigma(V_t)$ to be scaled down for both designs, as seen in Fig. 5.15, while planar MOSFETs remain at a fixed $\sigma(V_t)$ even for high-performance designs with α^{-1} scaling, as expected. Therefore, for LP designs (Fig. 5.16a), FinFETs reduce V_{min} to about 0.65 V for the BL and SRAMs and to about 0.46 V for DRAMs in the 11-nm

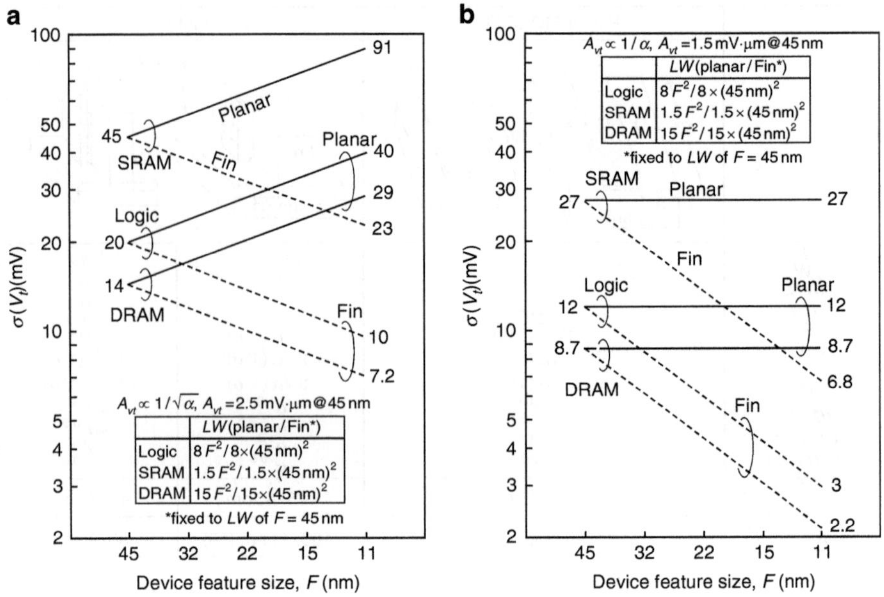

Fig. 5.15 Expected trends in $\sigma(V_t)$ for (a) low-power designs and (b) high-performance designs. Reproduced from [3] with permission; © 2010 IEEE

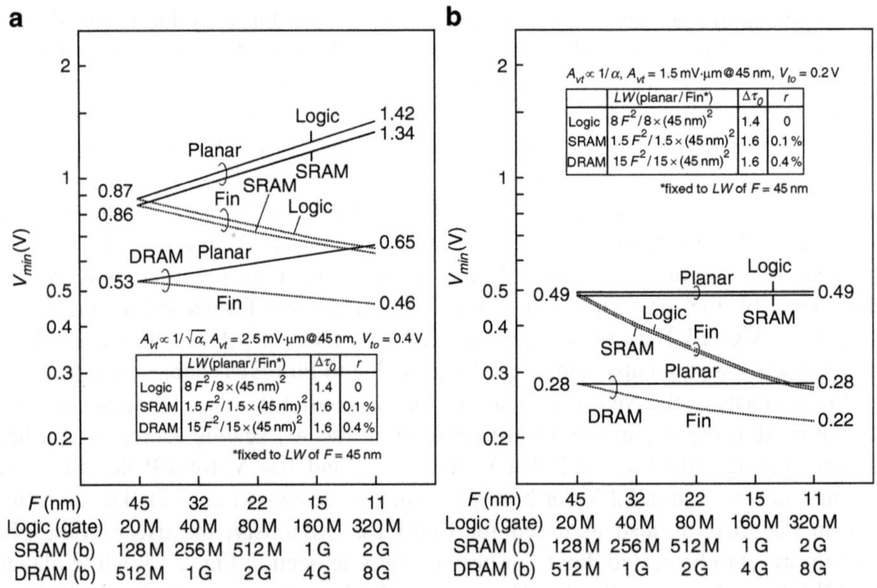

Fig. 5.16 Expected trends in V_{min} for (a) low-power designs and (b) high-performance designs. Reproduced from [3] with permission; © 2010 IEEE

generation. Such high V_{min}s result from using a high V_{t0} of 0.4 V. If V_{t0}-scalable and low-leakage circuits, which are described below, are used, the V_{min}s are effectively reduced to less than 0.5 V. Replacing SRAMs with DRAMs may also be effective to solve the high V_{min} problem of SRAMs. For high-performance designs (Fig. 5.16b), FinFETs reduce V_{min} to as low as about 0.27 V for the logic and SRAMs and 0.22 V for DRAMs.

5.5 Logic Circuits for Wider Margins

Low-V_{t0} circuits further reduce the V_{min} shown in Fig. 5.16. They are indispensable to reduce V_{min} not only for the logic block, but also for RAMs that include logic circuits in the periphery. Reducing V_{t0}, if possible, exceeding the lower limit shown in Fig. 5.5, directly reduces V_{min}. In addition, it further reduces V_{min} due to reduced $\sigma(V_t)$, as shown in Fig. 5.17 [13]. For example, if V_{t0} is reduced from 0.4 to 0 V, $\sigma(V_t)$ is reduced to more than a half. High-V_{t0} MOSFETs, however, must be added to cut leakage paths of low-V_{t0} MOSFETs, thus necessitating a high V_{DD} to ensure the high speed. Eventually, dual-V_{DD} and dual-V_{t0} circuits are indispensable. If low-V_{t0} MOSFETs operating at the low V_{DD} are used as much as possible in the dual-V_{t0} dual-V_{DD} circuits, they effectively reduce the V_{min} of the circuits. There have been two concepts to cut leakage paths of the low-V_{t0} MOSFETs. The first is the gate-source (G-S) reverse biasing of low-V_{t0} MOSFETs, which is further categorized as the G-S offset driving and the G-S differential driving. The second is the instantaneous activation of low-V_{t0} MOSFETs. Although the activation has recently been proposed for DRAM sense amplifiers, it is useful even for logic circuits. Note that the well-known gate-boosting of a high-V_{t0} MOSFET with a MOS capacitor also simultaneously achieves a high speed and low leakage, which was popular in the

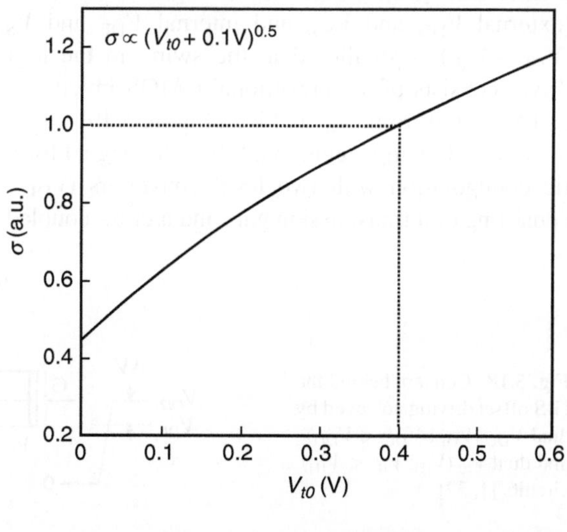

Fig. 5.17 $\sigma(V_t)$ vs. V_{t0} [13]

NMOS DRAM era in the 1970s [5]. This section describes these circuits and applications. Here, the threshold voltage of MOSFET V_t is again defined as the sum of V_t (nA/μm) and 0.3 V.

5.5.1 Gate-Source Offset Driving

One way to reduce the resultant leakage is to make the V_{t0} effectively high. Obtaining an effectively high V_{t0}, despite a low actual V_{t0}, can be achieved by using the G-S reverse biasing with the help of a high V_{DD} provided by a high-V_{DD} and high-V_{t0} circuit. The G-S reverse biasing is usually accomplished by the G-S offset driving of a low-V_{t0} MOSFET. Figure 5.18a shows the concept [32] behind the offset driving achieved by dual-V_{DD} (V_{DD} and $V_{DL} < V_{DD}$) and dual-V_{t0} (V_{tH} and $V_{tL} < V_{tH}$) circuits. It works with a large difference in V_{t0}, as exemplified by a high V_t (V_{tH}) of 0.4 V and a low V_t (V_{tL}) of zero. For example, reverse biasing is applied to a V_{tL}-PMOSFET during inactive periods with the help of a higher power supply V_{DD}. As a result, a sufficiently high V_{t0} (V_{teff}), despite a low-actual V_{tL}, is obtained, thereby reducing leakage during inactive periods. Even so, the gate-over-drive voltage (V_G) that determines the driving speed is maintained at a high level during active periods. Thus, V_{t0} can be scaled by adjusting the G-S bias. Note that even depletion (normally on) MOSFETs (i.e., D-MOSFETs) can be used, as explained later, as long as the MOSFET is cut by using a sufficiently high V_{DD}. The resultant circuit enables to minimize power dissipation with reduced voltage swing at the output if the low-V_{t0} MOSFETs are used at the output of inherently high-power circuits, such as the buffers, for driving heavy capacitive loads.

The basic scheme of the G-S offset driving was firstly proposed for the static bus architecture [32], as shown in Fig. 5.19a. It has two supply-voltage systems (external V_{DD} and V_{SS}, and internal V_{DL} and V_{SL}), and the bus signal swing ($V_{DL} - V_{SL}$) is smaller than the swing in the logic stage ($V_{DD} - V_{SS}$). The bus driver consists of a conventional CMOS inverter but with a low-V_t MOSTs (M_{P1} and M_{N1}) fed from V_{DL} and V_{SL}. Figure 5.19b shows a bus receiver which converts the reduced-swing signal to a full-swing signal for the logic stage. It has a symmetric configuration with two level converters ($V_{DL} \rightarrow V_{DD}$ and $V_{SL} \rightarrow V_{SS}$), each consisting of a transmission gate and a cross-coupled MOSFET pair. This efficiently

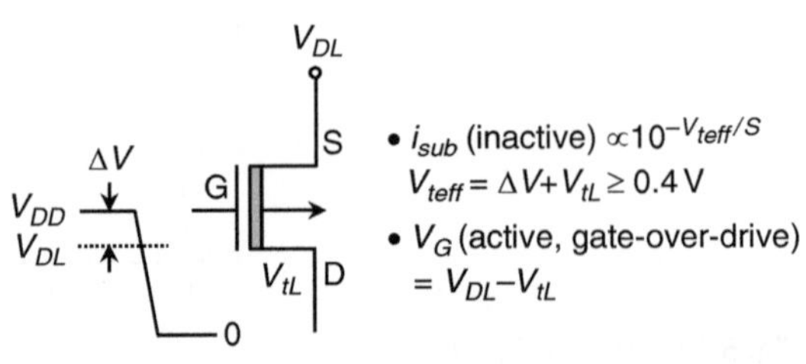

Fig. 5.18 Concept behind the G-S offset driving achieved by dual-V_{DD} (V_{DD}; $V_{DL} < V_{DD}$) and dual-V_{t0} (V_{tH}; $V_{tL} < V_{tH}$) circuits [1, 32]

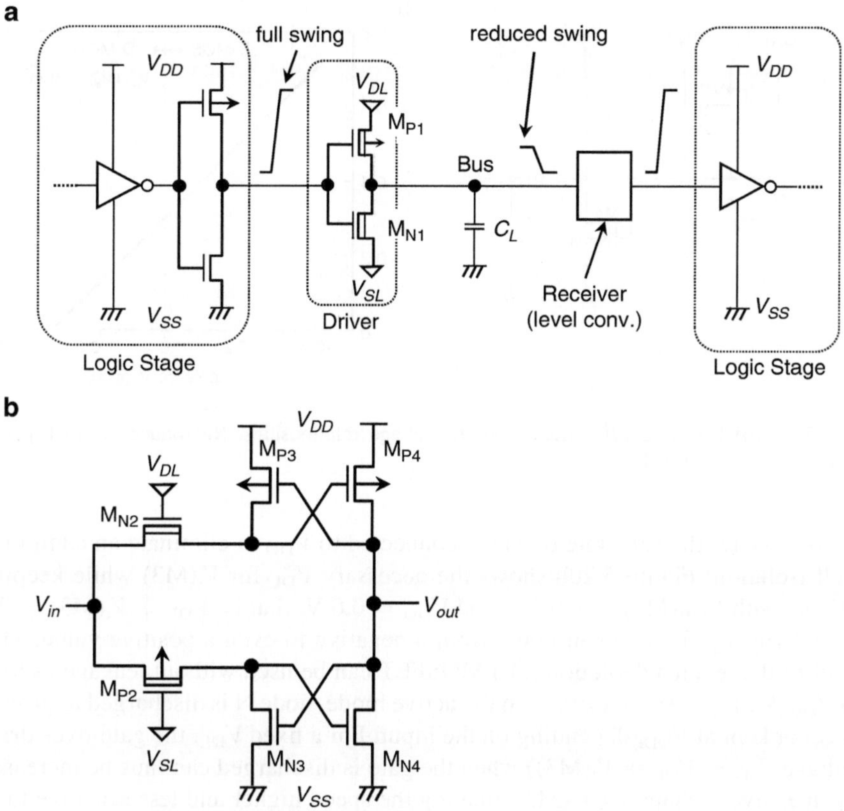

Fig. 5.19 (**a**) Bus architecture and (**b**) bus receiver. M_{P1}, M_{N1}, M_{P2}, and M_{N2}: low-V_{t0} MOSFETs. Reproduced from [32] with permission; © 2010 IEEE

converts the signal swing from V_{DL}–V_{SL} to V_{DD}–V_{SS} without increasing the standby current. The reduced voltage swing on the bus line with a heavy capacitance and the G-S offset driving are responsible for LP high-speed operations, and low-standby current.

The dynamic CMOS circuit using the concept, shown in Fig. 5.20a [1, 33], can also drive a heavily capacitive load at a low-voltage swing while minimizing the leakage and speed variation. The circuit consists of a V_{DL}-input NMOS M1 to gate an input (IN) with a V_{DL}-clock (CK1), a PMOS M2 to precharge node N to V_{DD} ($>V_{DL}$) at a V_{DD}-clock (CK2), and a V_{DL}-output E/D inverter (M3; D-MOS, M4; E-MOS). The threshold voltages of M2 and M4 are high enough to cut subthreshold currents, while those of M1 and M3 are variable. In the STB, M1 is cut off while M2 is turned on to precharge the gate of M3 at V_{DD}. The threshold voltage of M3 is thus effectively high because of $-(V_{DD} - V_{DL})$, so M3 is cut off if $V_{teff}(M3) \le -0.3$ V. Here, $V_{teff}(M3)$ is the resulting effective threshold voltage of $-(V_{DD} - V_{DL}) + V_t(M3)$, assuming that the actual threshold voltage of M3 is

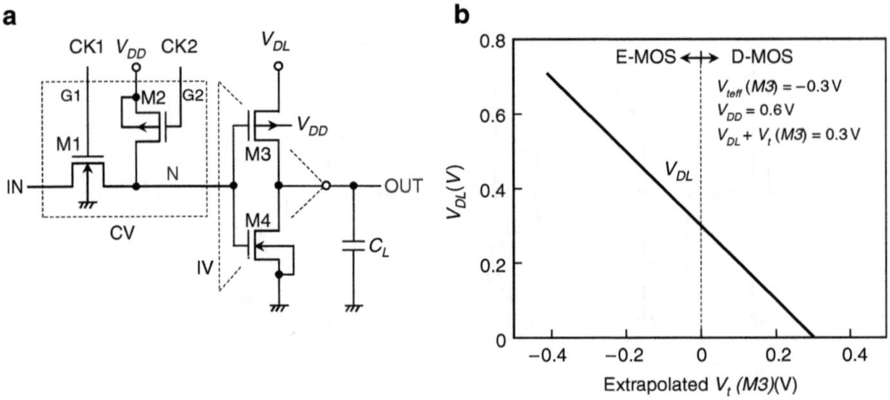

Fig. 5.20 (**a**) Dynamic E/D circuit and (**b**) voltage relationships. Reproduced from [1] with permission; © 2010 IEEE

V_t(M3). Here, the substrate of M3 is connected to V_{DD} to eliminate an additional well isolation. Figure 5.20b shows the necessary V_{DL} for V_t(M3) while keeping M3 off with $V_{teff}(M3) = -0.3$ V and $V_{DD} = 0.6$ V, that is, $V_{DL} + V_t(M3) = 0.3$ V. Obviously, V_t(M3) can range from a negative to even a positive value. This implies that even a depletion (D-) MOSFET can be used without leakage as long as $V_{teff}(M3) \leq -0.3$ V $(=V_{t0})$. In the active mode, node N is discharged to ground (V_{SS}) or kept at V_{DD}, depending on the input. For a fixed V_{DL}, the gate-over-drive voltage ($V_G = V_{DL} + V_t(M3)$) when the gate is discharged can thus be increased by intensively changing V_t(M3), making the speed higher and less sensitive to V_t variations ΔV_t. Moreover, for a fixed V_G, in other words, for a fixed speed, V_{DD} and thus power can be reduced, if the internal power of the circuit is much smaller than the output driving power.

Figure 5.21 illustrates the sensitivity of the input–output delay to V_t(M1) for $V_{DL} = 0.1$ V, $V_{DD} = 0.6$ V, and $V_t(M3) = 0.2$ V (i.e., D-MOS). The pulse width of CK1 must be wider than the low-input delay $\tau(L)$ and narrower than the high-input delay $\tau(H)$ to discriminate the input. For example, for an average $V_t(M1) = 0.3$ V and $\Delta V_t = \pm 50$ mV, the circuit successfully operates despite the ΔV_t, if the pulse width is controlled within point A (0.64 ns) and point B (1.12 ns). Figure 5.22 shows operating waveforms of the E/D circuit for $V_{t0}(M1) = 0.3$ V: the circuit operates even at $V_{DD} = 0.1$ V. For a low input after applying CK1 and CK2, node N (i.e., N(L)) is discharged to V_{SS} to drive the output, OUT(L), to V_{DL}. Even for a high input, however, it (i.e., N(H)) is gradually discharged from the V_{DD} level, although the output, OUT(H), is kept at the V_{SS} level. Note that V_t in Figs. 5.20–5.22 is the extrapolated V_t, in which 0.3 V is simply added to the constant current V_t (nA/μm), as mentioned earlier. Hence, M1 with an extrapolated V_t of 0.3 V can slowly discharge node N even at a 0.1-V high input and 0.1-V CK1. In any event, for a given V_G of 0.3 V, the output power is theoretically reduced to 1/36 that of the conventional 0.6-V static CMOS inverter with a $V_{t0} = 0.3$ V.

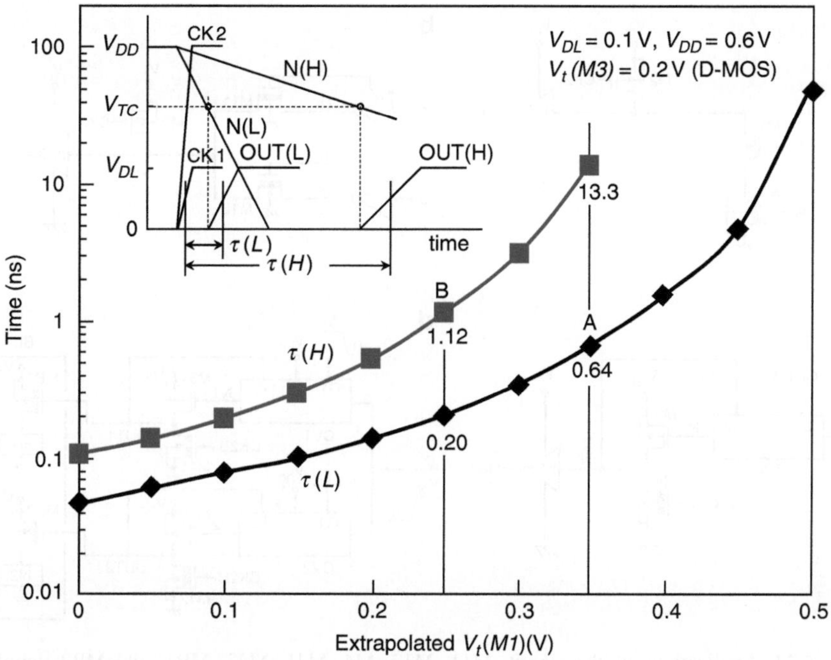

Fig. 5.21 Delay vs. V_t of the input MOS. 65-nm devices: $C_L = 4fF + 4MOSs$ ($W/L = 140/50$ nm); $V_t = 0.3$ V (M1), 0.3 V (M4), 0.2 V (M3, D-MOS), and −0.3 V (M2); $W/L = 140/50$ nm (M1, M2), 420/50 nm (M3), and 280/50 nm (M4). V_{TC}; logic threshold of the circuit. Reproduced from [1] with permission; © 2010 IEEE

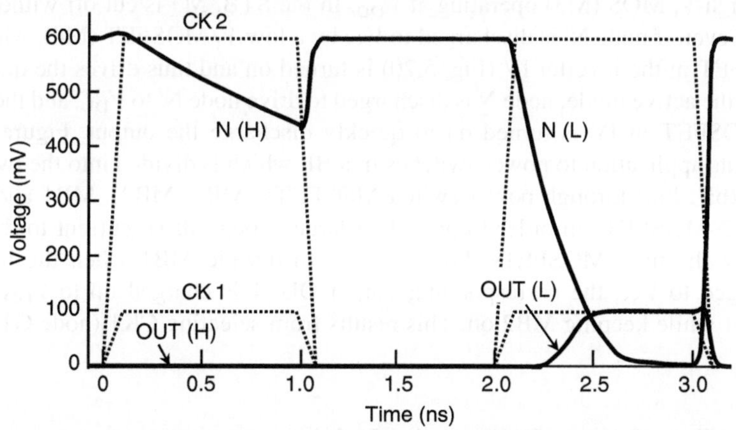

Fig. 5.22 Waveforms for $V_{DL} = 0.1$ V and $V_{DD} = 0.6$ V. All conditions are the same as those in Fig. 5.21. Reproduced from [1] with permission; © 2010 IEEE

The above concept is applicable to various dynamic logic circuits. Figure 5.23 shows applications to an AND gate (a) and an OR gate (b). Input MOSFETs (M11, M12) can accept a small enough V_t because the leakage current is reduced by the

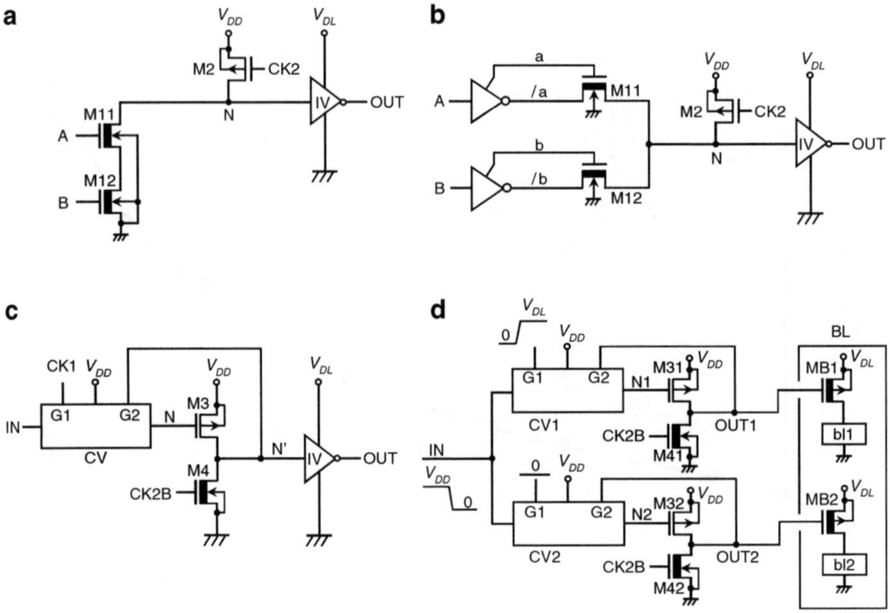

Fig. 5.23 Applications of the circuit. M11, M12, M4, M41, M42, MB1, and MB2: low-V_{t0} MOSFETs. Reproduced from [1] with permission; © 2010 IEEE

stacking effect for (a) [4, 5] or the G-S differential driving, described later, for (b). Figure 5.23c shows another version of the circuit using a feedback loop at node N′ and a high-V_t MOS (M3) operating at V_{DD}. In the STB, M3 is cut off without any leakage even if node N′ is discharged to V_{SS} by a low-V_t MOSFET (M4), while the PMOSFET in the inverter IV (Fig. 5.20) is turned on and thus drives the output to V_{DL}. In the active mode, node N is discharged to drive node N′ to V_{DD}, and the high-V_t NMOSFET in IV is turned on to quickly discharge the output. Figure 5.23d depicts its application to power switches in a BL which is divided into the two sub-blocks (bl1, bl2) through power switch MOSFETs (MB1, MB2). MB1 and MB2 can be D-MOSFETs in order to provide a large enough drive current to the sub-blocks with small MOSFETs. For example, to disable MB1 when the input is discharged to V_{SS}, the corresponding output OUT1 is charged up to V_{DD} to cut off MB1 while keeping MB2 on. This results from selecting CK1 (node G1).

5.5.2 Gate-Source Differential Driving

The G-S differential driving of a low-V_{t0} MOSFET, as shown in Fig. 5.24, is another G-S reverse biasing. It increases both the effective V_t, V_{teff}, and the gate-over-drive voltage, V_G, thus expanding operating V_{DL} and V_G regions, as shown in Fig. 5.25. The differential driving works within the region surrounded by line

	Differential Drive (A)	Conventional Drive (B)
Circuit	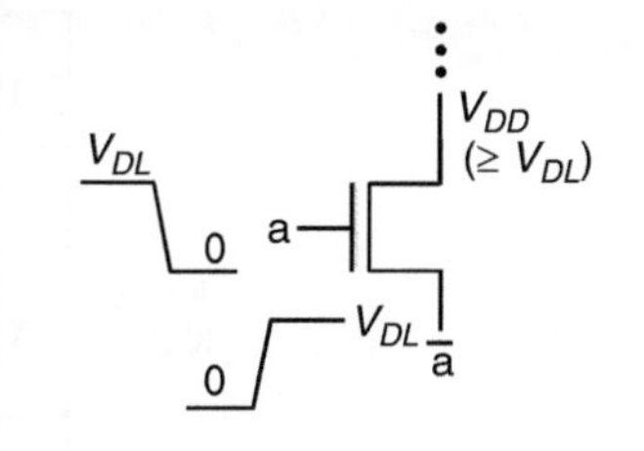	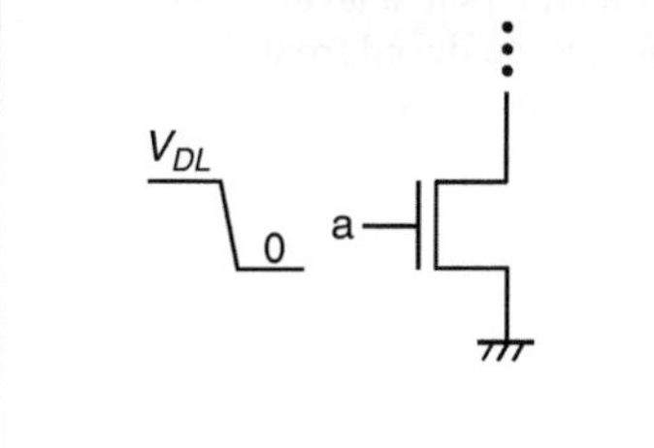
Effective V_t V_{teff}	$V_{DL} + V_t (> V_{t0})$	$V_t (> V_{t0})$
Gate-over-drive V_G	$V_{DL} - V_t$	$V_{DL} - V_t$

V_t: actual V_t, V_{t0}; lowest necessary V_t ($\cong$ 0.3 V)

Fig. 5.24 Differential drive (A) compared with conventional drive (B)

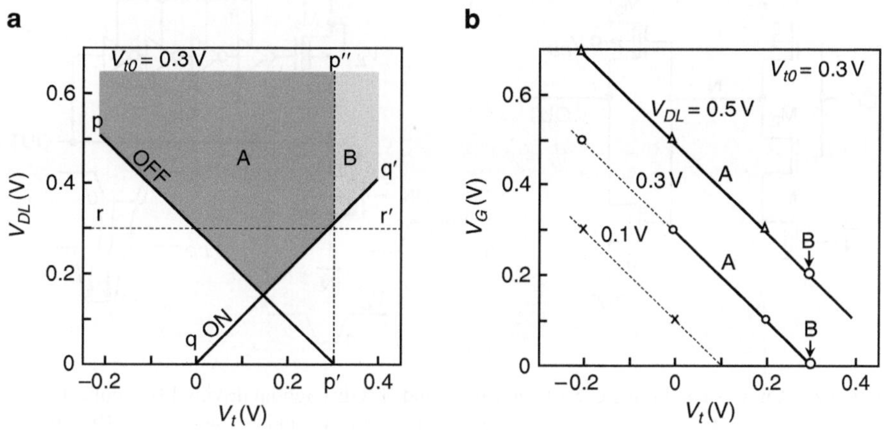

Fig. 5.25 Comparisons of (**a**) operating region and (**b**) gate-over-drive voltage between differential drive (A) and conventional drive (B)

pp′ (off-condition) and line qq′ (on-condition), while the conventional works within the region surrounded by line rr′ and line p′p″ (off-condition). Obviously, the differential driving enables to flexibly set V_{DL}, depending on the V_t. It also enables low-voltage operations with halved minimum V_{DL}. For example, for $V_{t0} = 0.3$ V, the minimum V_{DL} of the MOSFET (i.e., V_{min}) is as low as 0.15 V at $V_t = 0.15$ V, while 0.3 V for the conventional. Moreover, the differential driving achieves higher V_G. For example, the V_G at $V_{DL} = 0.3$ V is 0.3 V at $V_t = 0$ V for the differential driving, while 0 V for the conventional due to a fixed $V_t = 0.3$ V. Hence, the differential driving is suitable for low-voltage receivers or level shifters. Figure 5.26

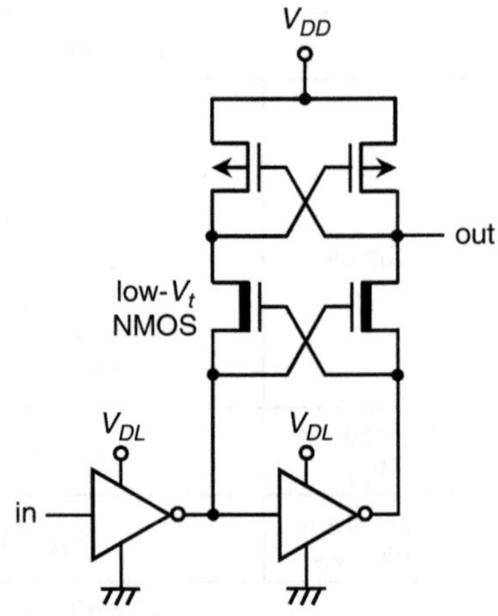

Fig. 5.26 Application of G-S differential driving of low-V_{t0} MOSFETs to a level shifter. Reproduced from [34] with permission; © 2010 IEEE

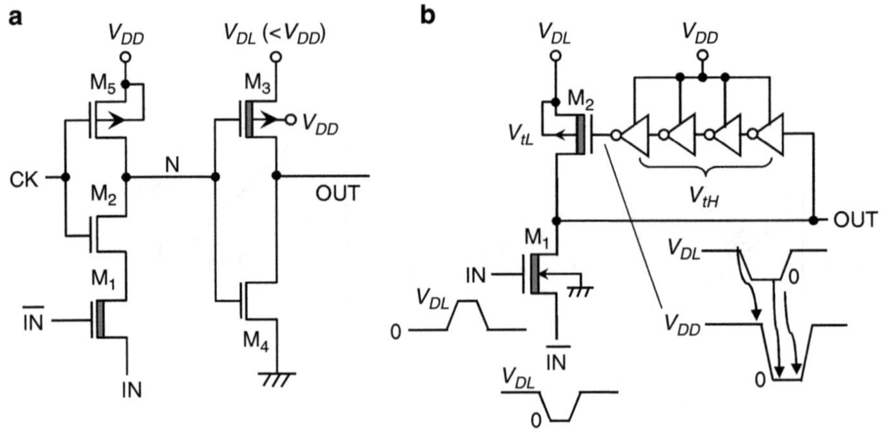

Fig. 5.27 Circuits combining G-S offset driving and G-S differential driving [3]. Applications to (**a**) inverter and (**b**) self-resetting circuit. M_1 and M_3 in (**a**), and M_1 and M_2 in (**b**): low-V_{t0} MOSFETs. Reproduced from [3] with permission; © 2010 IEEE

illustrates a static level shifter [34], in which each of low-V_{t0} cross-coupled MOSFETs is differentially driven.

5.5.3 Combined Driving

If the G-S offset and differential driving schemes are combined, excellent low-voltage circuits with wide voltage margins are realized, as exemplified by the circuits in Fig. 5.27 [35–38]. The circuit shown in (a) is a dynamic inverter [3, 33] with a V_{DD}-clock, CK, in which a low V_{t0} is assigned only for input detector M_1 and output driver M_3. The operation is almost same as that in Fig. 5.20. However, the voltage margin of the input circuit becomes wider due to differential driving. The circuit shown in (b) is a self-resetting circuit. When M_1 is on, the output goes to low, and the gate of M_2 becomes low, so M_2 drives the output to V_{DL}. After that, M_2 is turned off because a high-V_{DD} high-V_{t0} CMOS inverter chain drives the gate to V_{DD}, so the G-S is reverse biased by V_{DD}–V_{DL}. Although a leakage flows at the first stage inverter in the chain when output is at V_{DL}, it is small due to the small W.

5.5.4 Instantaneous Activation of Low-V_{t0} MOSFETs

The concept is vital even for logic circuits, although it was proposed for a low-voltage wide-margin DRAM sense amplifier, as discussed later. Figure 5.28 shows a low-V_{t0} circuit backed up by a high-V_{t0} circuit. Once the low-V_{t0} circuit has detected a low-voltage input signal, it is turned off to cut the leakage path. After that, the detected signal is held thereafter in the high-V_{t0} circuit without leakage. The leakage of the whole circuit is thus small because a large leakage flows only instantaneously in the low-V_{t0} circuit.

5.5.5 Gate Boosting of High-V_{t0} MOSFETs

The gate boosting raises the gate voltage of a high-V_t MOSFET to eliminate the V_t-drop and increase the gate-over-drive voltage even under a low drain voltage, which had been popular in the NMOS DRAM era [5]. Figure 5.29 shows such a

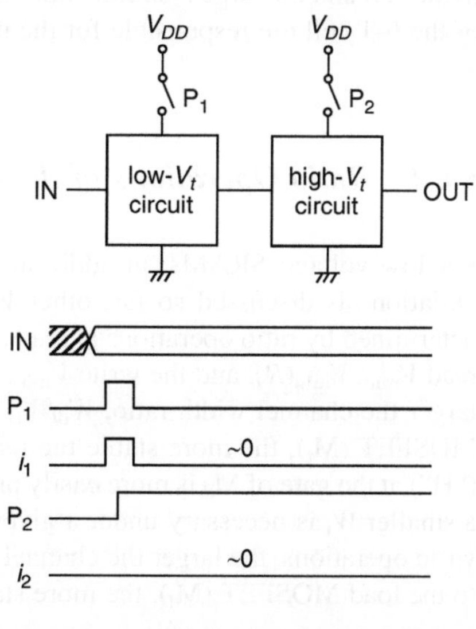

Fig. 5.28 Concept behind instantaneous activation of low-V_t circuit backed up by high-V_t circuit. i_1, i_2: leakage current

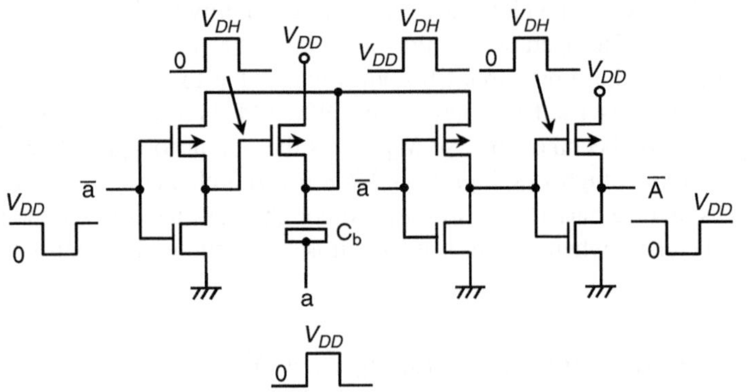

Fig. 5.29 Node boosting with a MOS capacitor C_b [39]. V_{DH}: boosted voltage

circuit [39] that can operate even at $V_{DD} = 0.3$ V by boosting the gate to about 0.6 V ($=V_{DH}$) with a MOS capacitor.

5.6 SRAMs for Wider Margins

Recent research on high-speed low-voltage 6-T SRAMs has focused on widening the voltage margin at a fixed operating voltage of around 1 V rather than reducing V_{min} and thus V_{DD}. This is because even at a V_{DD} of 1 V, despite designers' need for further reducing the V_{DD}, the voltage margin has been rapidly reduced with device scaling, and thus margin errors have increased. In addition to the smallest MOSFET in the cell and the largest circuit count in a chip, as discussed so far, ratio operations of the 6-T cell are responsible for the narrower margin.

5.6.1 Ratio Operations of the 6-T Cell

For low-voltage SRAMs, in addition to the V_{min} defined by the tolerable speed variation, as discussed so far, other V_{min}s defined by the voltage margin that is determined by ratio operations at read and write must also be reduced. They are the read V_{min}, $V_{min}(R)$, and the write V_{min}, $V_{min}(W)$. At read operations (Fig. 5.30), the larger the channel width ratio, W_d/W_t, of the driver MOSFET (M_d) to the transfer MOSFET (M_t), the more stable the read operation. This is because the high state ("H") at the gate of M_d is more easily preserved when M_t is turned on, implying that a smaller W_t is necessary under a given cell size for lower $V_{min}(R)$. In contrast, at write operations, the larger the channel width ratio, W_t/W_l, of the transfer MOSFET to the load MOSFET (M_l), the more stable the write operation. This is because the

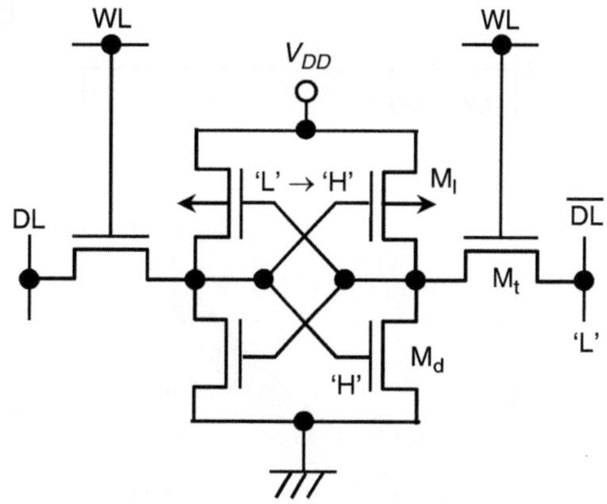

Fig. 5.30 Ratio operations of 6-T cell

low state ("L") at the gate of M_l more easily flips to "H" when "L" is applied from the dataline, implying that a larger W_t is necessary for lower $V_{min}(W)$. These contradictory requirements for M_t make SRAM designs complicated. Unfortunately, $V_{min}(R)$ and $V_{min}(W)$ are strongly influenced by V_t-variations of MOSFETs and imbalances between each pair of MOSFETs in the cell, despite a lithographically symmetric cell layout being used [4]. Although relationships between the V_{min} defined by the timing margin and the V_{min}s (i.e., $V_{min}(R)$ and $V_{min}(W)$) defined by the voltage margins remain unclarified, many attempts have been made to reduce their V_{min}s.

5.6.2 *Shortening of Datalines and Up-Sizing of the 6-T Cell*

Shortening the dataline [40] reduces the V_{min} of the 6-T cell because a large speed variation $\Delta\tau_0$ is allowed. For example, if the dataline (bit-line) length is halved to increase $\Delta\tau_0$ from 1.6 to 3.6, V_{min} is reduced, as shown in Fig. 5.31a. Up-sizing MOSFETs in the 6-T cell with the largest MOSFET possible [17, 41] also reduces V_{min} with reduced $\sigma(V_t)$. For example, if the channel lengths of all MOSFETs are scaled down while keeping the channel widths fixed, such as in the 90-nm generation (where $LW \propto F$ with W fixed at 90 nm; Fig. 5.31b), the increase in V_{min} can be suppressed. In contrast, with conventional scaling (i.e., $LW \propto F^2$), V_{min} rapidly increases as F decreases. The cell size (Fig. 5.32) of the W-fixed approach, however, is more gradually reduced since all Ws in the cell are fixed at each generation. Thus, the size becomes equal to that of an 8-T cell having a size of 156–185F^2 in the 45-nm generation, while the conventional scaling reduces the size more rapidly (i.e., with 120F^2). In practice, the sizes of MOSFETs in a 6-T cell can be adjusted between the two approaches, so the V_{min} is between about 0.6 and 1 V in the 32-nm generation for A_{vt} = 2.5 mV·µm, as seen in Fig. 5.31b.

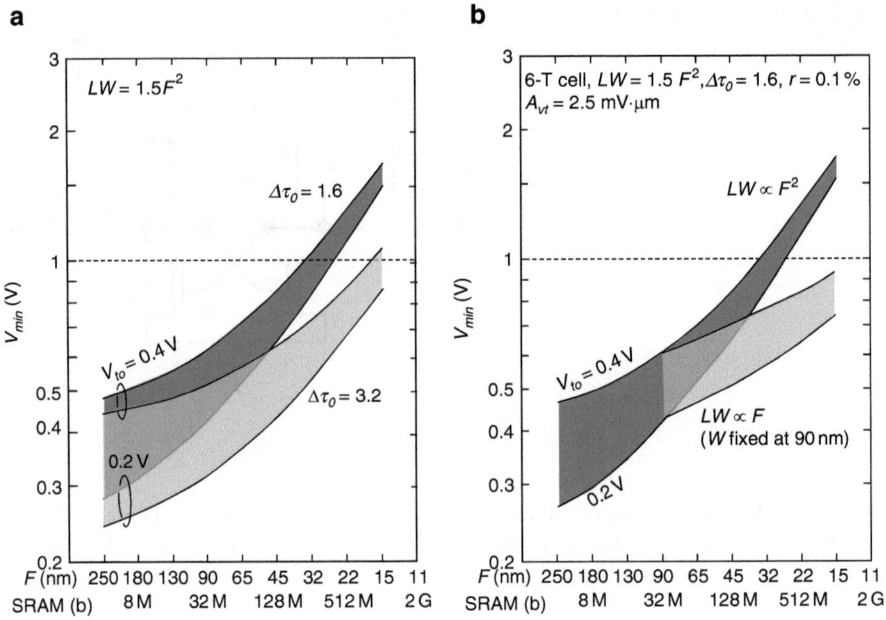

Fig. 5.31 $V_{\min}$ of 6-T cell: (**a**) shortening dataline and (**b**) up-sizing [11]

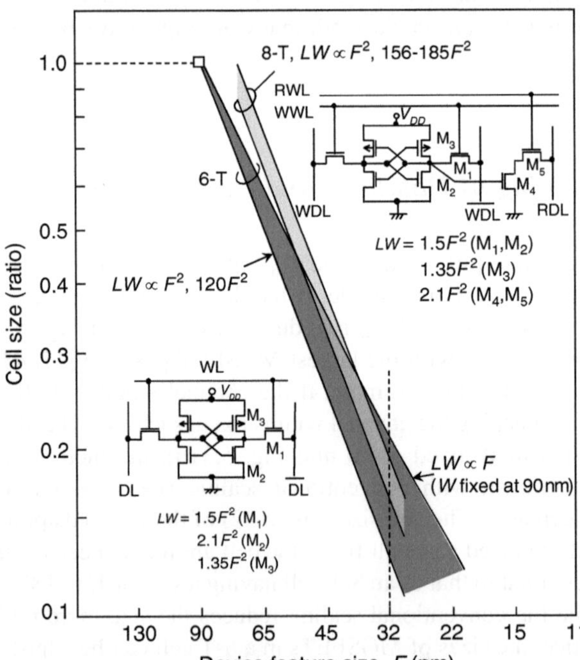

Fig. 5.32 Cell size of 6-T and 8-T cells. Reproduced from [3] with permission; © 2010 IEEE

5.6.3 Power Managements of the 6-T Cell

Power managements of the 6-T cell also reduce the V_{min}s. Figure 5.33 illustrates practical power managements applied to 6-T cells. The one shown in (a) [42] combines a low-V_t (V_{tL}) transfer MOSFET, a negative wordline, and reduced dataline voltage, V_{DL}. It reduces $V_{min}(W)$ while keeping the leakage low during nonselected periods. $V_{min}(W)$ and $V_{min}(R)$ are lower than those of the combination of a high-V_t (V_{tH}) transfer MOSFET and boosted wordline [43]. This is because the low V_t reduces the $\sigma(V_t)$ of conventional MOSFETs. In this scheme, as the data (bit) line voltage can be scaled in accordance with MOSFET scaling in the peripheral circuits, high density and low power are achieved for dataline-relevant circuits. Power control of PMOSFET loads (Fig. 5.33b) [14, 15] to increase load impedance during write periods improves the write margin, namely, reduces $V_{min}(W)$. Dynamic power control of the driver NMOSFET [17, 41, 44, 45] or load PMOSFET [46, 47] reduces the V_t in active mode (ACT) while reducing leakage in STB with increased V_t ($=\delta V_t$) due to the body bias effects. Figure 5.34 shows the dynamic control of the common source line of the cells [41, 45] with the switched-source impedance (SSI). In the STB, the leakage currents of the cells flow into the SSI, which raises the source to δ, thereby reducing the currents by the increased V_ts of M_d and M_t due to body effects and by a G-S reverse bias to M_t. Here, the increased V_ts must be equal to V_{t0} to ensure a low enough standby leakage. In the active mode, the source goes down to 0 V, so the V_ts becomes lower than V_{t0}, which reduces V_{min}.

A reduced wordline voltage scheme in accordance with the V_{t0} of the transfer MOSFET [16, 48, 49] has also been proposed to widen the read margin. Figure 5.35 depicts $V_{min}(R)$ and $V_{min}(W)$ vs. word voltage (V_{WL}) for a typical design, in which $V_{min}(R)$ is higher than $V_{min}(W)$ at $V_{WL} = V_{DD}$. $V_{min}(R)$ decreases when V_{WL} and

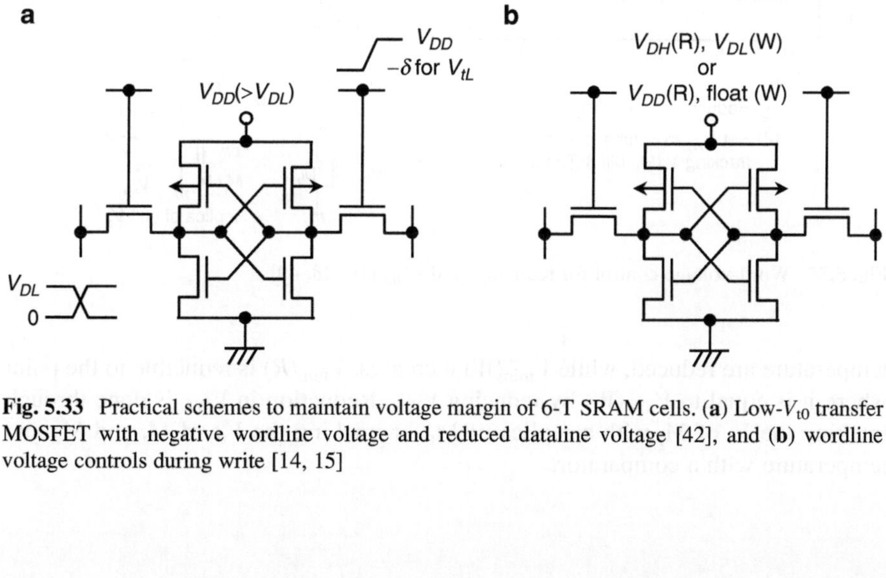

Fig. 5.33 Practical schemes to maintain voltage margin of 6-T SRAM cells. (**a**) Low-V_{t0} transfer MOSFET with negative wordline voltage and reduced dataline voltage [42], and (**b**) wordline voltage controls during write [14, 15]

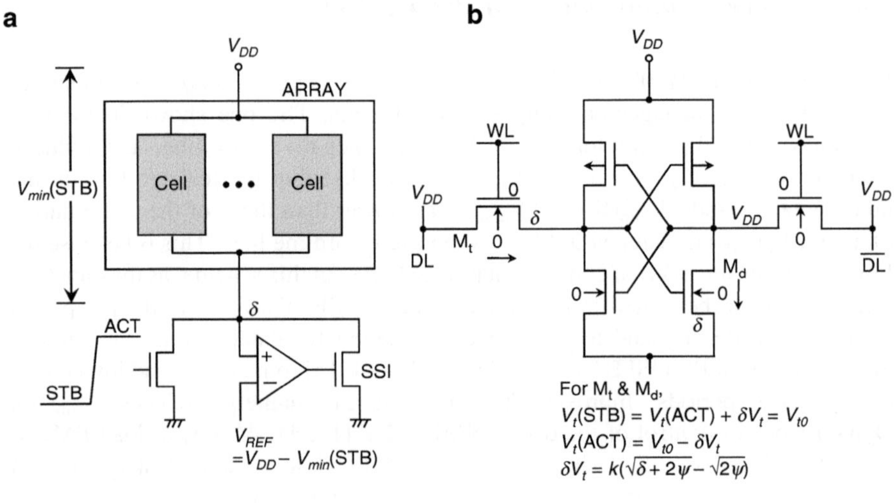

Fig. 5.34 Dynamic control of common source line of cells [41, 45]

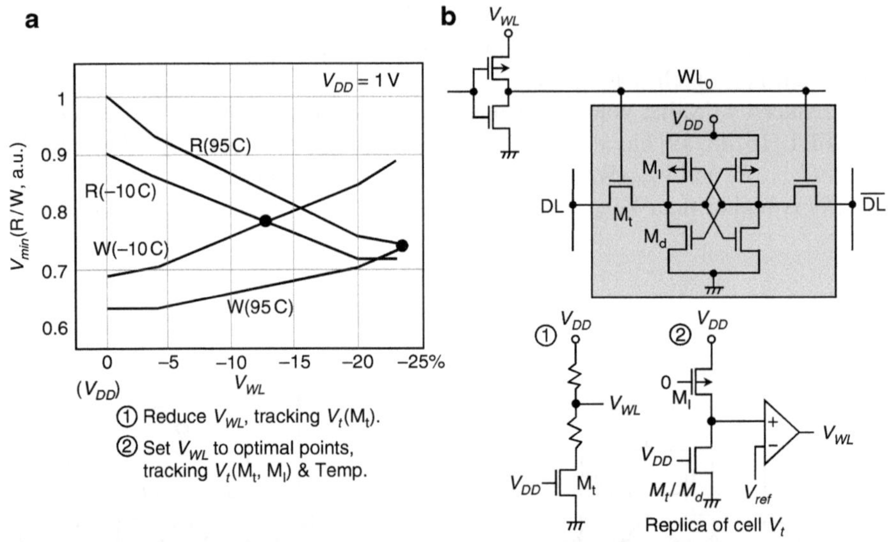

Fig. 5.35 Word-voltage control for reducing read $V_{\min}$ [16, 48, 49]

temperature are reduced, while $V_{\min}(W)$ increases. $V_{\min}(R)$ is reducible to the point where it is equal to $V_{\min}(W)$ by reducing V_{WL}. Reduction in V_{WL} is done through tracking the V_{t} of $\mathrm{M_t}$ with a replica of $\mathrm{M_t}$, or tracking the V_{t}s of $\mathrm{M_t}$ and $\mathrm{M_d}$ and temperature with a comparator.

5.6.4 The 8-T Cell

Different cell configurations such as 8-T cells [50] have also been proposed. Figure 5.36 shows the 8-T cell compared with the 6-T cell. The 8-T cell dramatically reduces $V_{min}(R)$ and $V_{min}(W)$ as well as the V_{min} defined previously by a timing margin (i.e., $V_{min}(\Delta\tau)$). The 8-T cell can perform the read operation using separate read path, M_r and M_s, without the ratio operation of M_t and M_d, which enables large W_r and W_s. $V_{min}(R)$ and $V_{min}(\Delta\tau)$ are thus reduced. The write operation can be done in the same manner as the 6-T cell, but a larger W_t is acceptable due to no restriction from the read operation, unlike the 6-T cell. $V_{min}(W)$ is thus reduced. This is correct as far as the selected cell is concerned. The widened W_t, however, poses a "half-select" problem for other memory cells along the selected write wordline: while one selected cell is written, the remaining half-selected cells are read (Fig. 5.37). Unfortunately, the cells may cause unstable – and in some cases, destructive – read operations due to reduced W_d/W_t. This means that the 8-T cell has limited applications to a wide-bit organization, in which all cells are simultaneously written.

The investigation suggests that multiple cell sizes and types combined with multi-V_{DD} operation on a chip are feasible, depending on the length of the dataline and the required memory chip capacity. For example, for a small-capacity SRAM, in which overhead due to the use of ECC is intolerable but a larger cell size is tolerable, up-sizing of MOSFETs in the cell enables low-V_{DD} operation. For a large-capacity SRAM necessitating a small cell size, repair techniques and/or a dedicated high-voltage supply are a viable solution. However, even if V_{DD} can be managed so that it remains at about 1 V even in 45–32-nm generations, it will still continue increasing, especially for conventional scaling aiming at higher density, as long as conventional MOSFETs are used. That is why small-A_{vt} MOSFETs as mentioned previously are indispensable, especially for such SRAMs.

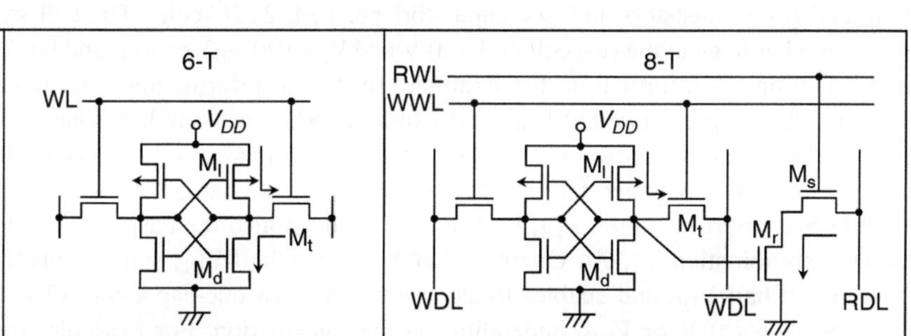

	6-T	8-T
(R)	The larger the W_d/W_t, the more stable the operation. → Smaller W_t	Using separate R-path (M_r, M_s) without ratio operation → Large W_r & W_s
(W)	The larger the W_t/W_l, the more stable the operation. → Larger W_t	Same as 6-T, but W_t can be widened due to no restriction from R-operation.

Fig. 5.36 Comparisons of voltage margins of selected 6-T and 8-T cells [50]

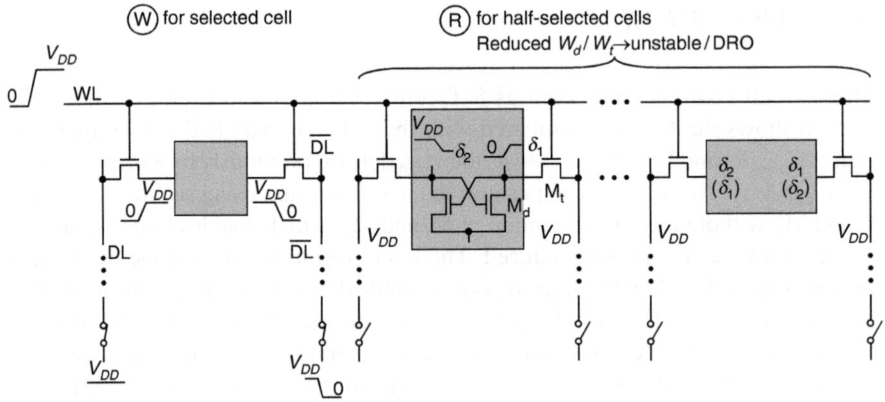

Fig. 5.37 Half-select problem of the 8-T cell

5.7 DRAMs for Wider Margins

5.7.1 Sensing Schemes

The V_{min} of the DRAM block is lowest, as discussed previously. This is correct as far as the V_{DD}-sensing is used. In practice, however, the half-V_{DD} sensing, instead of the V_{DD} sensing, has been used in products. The reason why the V_{DD} sensing has not been used so far despite the lower V_{min} is that it entails larger cell and higher power dissipation [4, 5], as explained below.

Figure 5.38 depicts two practical sensing schemes: the V_{DD} sensing and the half-V_{DD} sensing [4, 5]. The V_{DD} sensing precharges datalines at V_{DD} and necessitates a twin cell (two transistors and two capacitors per cell, 2T2C cell). The cell stores a differential voltage at the two cell nodes, 0 V and V_{DD} (0/V_{DD}), or V_{DD} and 0 (V_{DD}/0), depending on the information. For example, for 0/V_{DD} information, after datalines, DL and $\overline{\text{DL}}$, are precharged at V_{DD}, and wordline WL is activated, a negative signal component v_S is developed on DL, while no signal component on $\overline{\text{DL}}$ because of no voltage difference between $\overline{\text{DL}}$ and the cell node. The signal is then amplified by a CMOS SA, referring to the V_{DD} level on $\overline{\text{DL}}$. This is also the case for an opposite voltage combination V_{DD}/0. On the other hand, the half-V_{DD} sensing precharges datalines at half-V_{DD} and enables to use a one-transistor one-capacitor (1T1C) cell. The cell stores 0 V or V_{DD}, depending on the information. For example, for 0-V information, a negative signal component v_S is developed on DL and then amplified, referring to the half-V_{DD} level on $\overline{\text{DL}}$. For V_{DD} stored voltage, a positive signal component is developed and then amplified, referring to the half-V_{DD} level. It is obvious that the V_{DD} sensing doubles the cell size and dataline charging power (caused by the voltage swing on datalines) of the half-V_{DD} sensing. However, it can halve the V_{min} of the half-V_{DD} sensing, as explained later. Therefore, if a new half-V_{DD} sensing that enables the same V_{min} as the V_{DD} sensing is developed, while maintaining

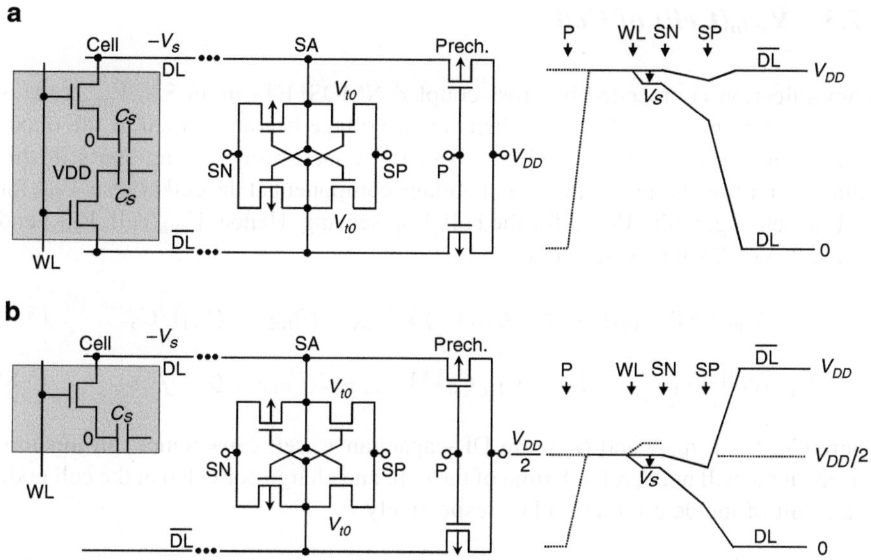

Fig. 5.38 Practical sensing schemes: (**a**) V_{DD} sensing, and (**b**) half-V_{DD} sensing [4, 5]

advantages of the small cell and low power, it enables low-cost LP DRAMs with wide margins even at low V_{DD}. Moreover, if combined with a logic-process compatible cell, cheaper embedded DRAMs that may enable to replace SRAMs are realized. The challenge to such a cell is to minimize the necessary cell capacitance C_S through reducing offset voltage of SA, leakage and SER of cells, and dataline parasitic capacitance C_D. The details are in what follows.

5.7.2 V_{min}(SA) of Sense Amplifier

For DRAMs, the V_{min} is equal to the higher of the two V_{min} values: the V_{min} of sense amplifiers (SAs), V_{min}(SA), determined by the necessary timing margin $\Delta\tau_0$, as defined previously, and the V_{min} of cells, V_{min}(cell), determined by signal-to-noise ratio (S/N) of the cell. In the past, V_{min}(SA) has eventually been higher than V_{min}(cell). The V_{min}(SA) of the V_{DD} sensing, V_{min}(SA, V_{DD}), and the V_{min}(SA) of the half-V_{DD} sensing, V_{min}(SA, V_{DD}/2), are expressed as

$$V_{min}(\text{SA}, V_{DD}) = V_{t0} + (1+\gamma)\Delta V_{tmax}, \tag{5.5}$$

$$V_{min}(\text{SA}, V_{DD}/2) = 2\{V_{t0} + (1+\gamma)\Delta V_{tmax}\}, \tag{5.6}$$

where V_{min}(SA, V_{DD}/2) is given by doubling V_{min}(SA, V_{DD}) because the V_{DD} in (5.2) is regarded as V_{DD}/2.

5.7.3 V_{min}(Cell) of Cell

If amplification is preceded by cross-coupled NMOSFETs in an SA, V_{min}(cell) is approximately given as the V_{DD} when signal voltage becomes equal to the maximum variation in V_t, ΔV_{tmax}. This is because a successful sensing starts in this voltage condition. Note that the signal voltage component at the cell node is V_{DD} for the V_{DD} sensing, while $V_{DD}/2$ for the half-V_{DD} sensing. Hence, V_{min}(cell, V_{DD}) and V_{min}(cell, $V_{DD}/2$) are expressed as

$$V_{\min}(\text{cell}, V_{DD}) \cong (1 + C_D/C_S)\Delta V_{tmax} + (it_{REF} + Q_{col})/C_S, \quad (5.7)$$

$$V_{\min}(\text{cell}, V_{DD}/2) \cong 2(1 + C_D/C_S)\Delta V_{tmax} + 2(it_{REF} + Q_{col})/C_S, \quad (5.8)$$

where C_D, C_S, i, t_{REF}, and Q_{col}, are DL-capacitance, cell capacitance, pn-junction current at the cell node, refresh time of the cell, and charge collected at the cell node as a result of incident radiation [1], respectively.

5.7.4 Comparison Between V_{min}(SA) and V_{min}(Cell)

In the past, to fully utilize the advantages of the half-V_{DD} sensing V_{min}(SA, $V_{DD}/2$) and V_{min}(cell, $V_{DD}/2$), despite inherently high values, have been reduced to practical levels. Consequently, the V_{min} of DRAMs has been equal to V_{min}(SA, $V_{DD}/2$) as a result of further reducing V_{min}(cell, $V_{DD}/2$). This subsection thus focuses on the half-V_{DD} sensing, and compares V_{min}(SA, $V_{DD}/2$) and V_{min}(cell, $V_{DD}/2$), simply expressed after this as V_{min}(SA) and V_{min}(cell). Here, it is assumed that $m = 5.5$ (no repair), $V_{t0} = 0.4$ V, $\gamma = 2.09$, and the sum of it_{REF} and Q_{col} is neglected for comprehensive explanation. C_D/C_S is taken as a parameter, because it has gradually been reduced from about 10 to below 5 with reducing C_D and keeping C_S almost constant. Figure 5.39a shows their expected trends for the conventional MOSFET with $A_{vt} = 4.2$ mV μm achieved in the 130-nm generation. Before the 130-nm generation, C_D/C_S was about 10, and V_{min}(SA) was thus higher than V_{min}(cell). After that, V_{min}(cell) is expected to rapidly increase due to the ever-larger ΔV_{tmax} and surpass V_{min}(SA), calling for reducing V_{min}(cell) to a level lower than V_{min}(SA) with reducing C_D/C_S. With $C_D/C_S = 5$, V_{min}(cell) is drastically reduced, so that it is lower than V_{min}(SA). Even so, 1-V operation is expected to be impossible over device generations. Figure 5.39b shows the trends for a more advanced MOSFET with $A_{vt} = 1.5$ mV μm achieved in the 45-nm generation. Obviously, V_{min}(cell) is particularly reduced mainly due to reduced ΔV_{tmax}, so the V_{min} of DRAMs is determined by V_{min}(SA) at every generation. Due to an extremely high level of V_{min}(SA), however, even 1-V operation is impossible despite such an advanced MOSFET. This problem, however, is resolved if V_{min}(SA) is halved, as illustrated with dotted line, by halving the sum of V_{t0} and

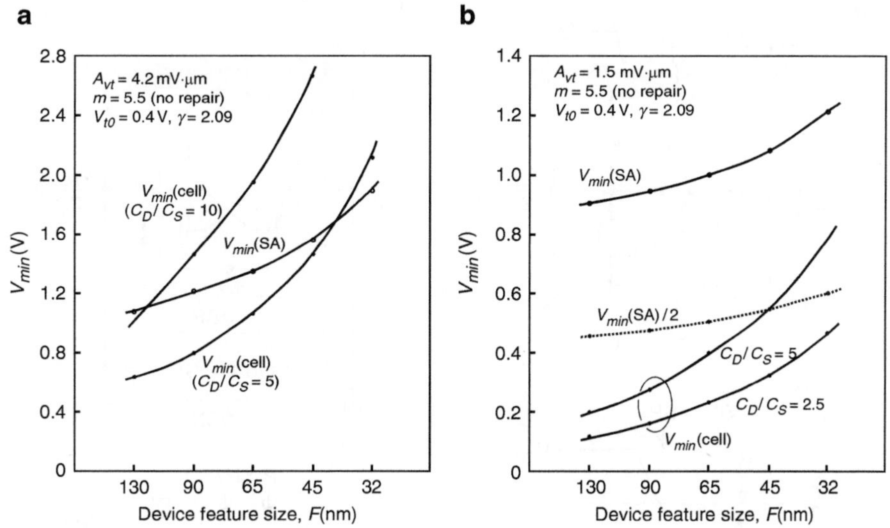

Fig. 5.39 Trends in V_{min} (SA) and V_{min} (cell) of the half-V_{DD} sensing for (**a**) A_{vt} = 4.2 mV·µm and (**b**) A_{vt} = 1.5 mV·µm

$(1 + \gamma)\Delta V_{tmax}$ in (5.6). Obviously, the halved V_{min}(SA) is equal to the V_{min}(SA) of the V_{DD}-sensing. That is why reducing the V_{t0} as well as ΔV_{tmax} of MOSFETs in SAs is critical for reducing the V_{min} of DRAMs. Here, if V_{t0} is reduced from 0.4 to 0 V, $\sigma(V_t)$ is reduced to more than a half (Fig. 5.17), as mentioned previously. Note that if repair techniques are adopted, both V_{min}(SA) and V_{min}(cell) are reduced.

5.7.5 Low-V_{t0} Sense Amplifier

A challenge to low-V_{min} DRAMs is to develop a low V_{t0} but low-leakage SA suitable for the half-V_{DD} sensing, as mentioned before. Figure 5.40 shows such an SA [5, 51–53] compared with the conventional high-V_{t0} cross-coupled CMOS SA. The conventional performs two functions, sensing and data holding, with one high-V_{t0} SA, while the new SA performs the two functions with two SAs [51]: a low-V_{t0} instantaneously activated preamplifier PA (M_P; low-V_{t0} NMOS) for low-voltage and low-leakage sensing, and the conventional cross-coupled high-V_{t0} CMOS SA (Fig. 5.38) for low-leakage data holding. After amplifying the signal to some extent by applying a short pulse, P, the low-V_{t0} PA is turned off to cut the leakage path at M_P. The high-V_{t0} SA is then activated to latch and hold the amplified signal. In this manner, low-voltage sensing and low-leakage data holding are simultaneously performed. To be more precise, the PA stops amplification when DL drops to $V_{t0}(M_P)$, enabling the signal to be finally amplified to $V_{t0}(M_P)$. For the high-V_{t0} SA in order to

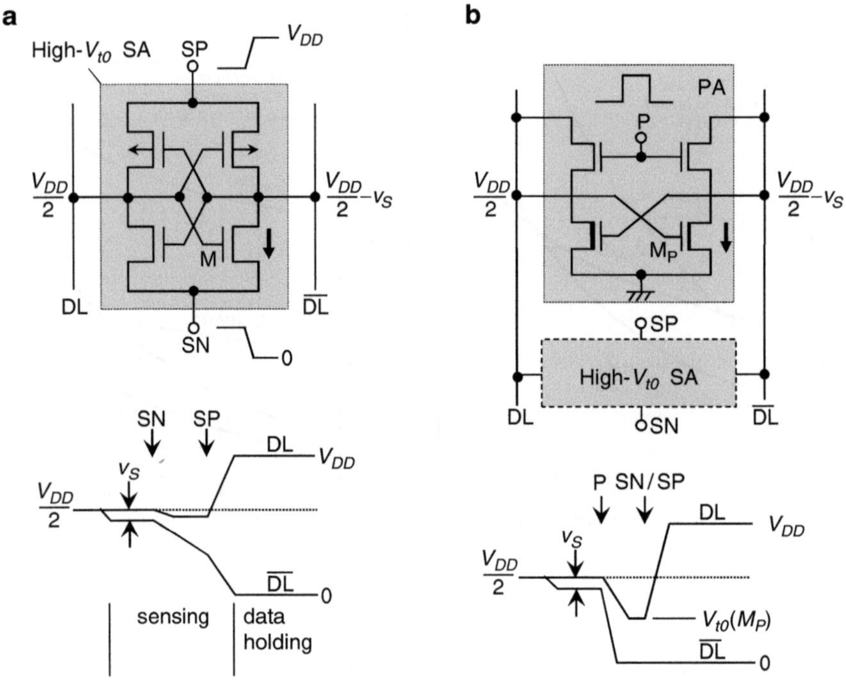

Fig. 5.40 (**a**) Conventional sense amplifier and (**b**) dual sense amplifier [5, 51–53]. Mp: low-V_{t0} MOSFET. Reproduced from [11] with permission; © 2010 IEICE

successfully latch the amplified signal, $V_{t0}(M_P)$ must be higher than the offset voltage of cross-coupled PMOSFETs in the high-V_{t0} SA (because NMOSFETs are off at the initial stage of latching) that is usually less than 0.2 V. Therefore, $V_{t0}(M_P)$ can be less than 0.2 V. Hence, the half V_{DD} can be reduced to a level close to such a low $V_{t0}(M_P)$, where the PA turns on. The area penalty has been reported to be 2.2% for a 128-Mb DRAM [51].

5.7.6 FD-SOI Cells

In practice, detrimental effects of the leakage current and radiation incident (i.e., it_{REF} and Q_{col}) are not negligible and thus worsen the voltage margin of the cell. They must be resolved by means of innovative device structures. If the devices can reduce dataline capacitance C_D, they reduce V_{min}(cell). A promising candidate is DRAM cells using FD SOI structures. In fact, FD-SOI structures are beneficial most to DRAMs, as shown in Fig. 5.41 [24, 54], because a small and thus simple capacitor is realized. Such a capacitor is enabled by reduced offset of an SA due to reduced V_t-variations, small pn-junction leakage, negligible SER due to ultrathin

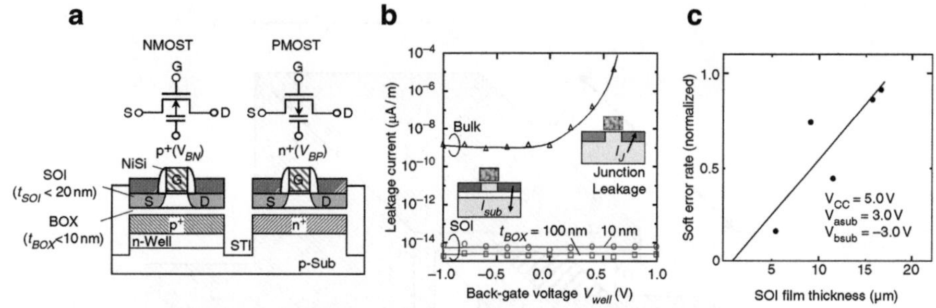

Fig. 5.41 (**a**) Structures of planar UT-BOX FD-SOI [24] and (**b**) leakage currents compared with bulk structure [24], and (**c**) soft error rates of a FD-SOI [54]

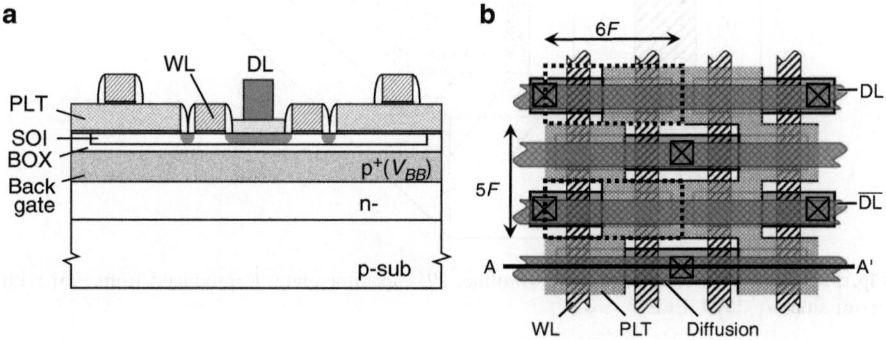

Fig. 5.42 (**a**) Cross section and (**b**) layout of a UT-BOX FD-SOI twin-cell [55]. Reproduced from [1] with permission; © 2010 IEEE

silicon film, and small dataline capacitance due to its small parasitic capacitance. Even wordline voltage is reducible due to no body effect.

Figure 5.42 illustrates a logic-process-compatible planar-FD-SOI twin DRAM cell [55]. The V_t of the cell transistor is adjusted with ion implantation under the BOX and/or well bias voltage (i.e., back-gate voltage V_{BB}). To keep the junction leakage in the MOSFET extremely low, no implant is added to the store node. The SER is negligible due to a small collection area of the SOI. Each dataline is shielded with quiescent adjacent datalines to avoid the DL–DL coupling noise. The size of the twin cell for a 0.5-fF C_S is $30F^2$, which is less than one-fourth that of the 6-T SRAM-cell. Figure 5.43 shows a logic-process-compatible 10-F^2 FinFET DRAM cell designed using a side-wall process [26, 29]. It adopts the UT-BOX structure to control the V_t by applying V_{BB}. The cell capacitor uses a FinFET structure to double the capacitance (Fig. 5.13a), while the transfer MOSFET uses another structure to reduce the gate capacitance that works as a parasitic capacitance when the cell is read (Fig. 5.13c).

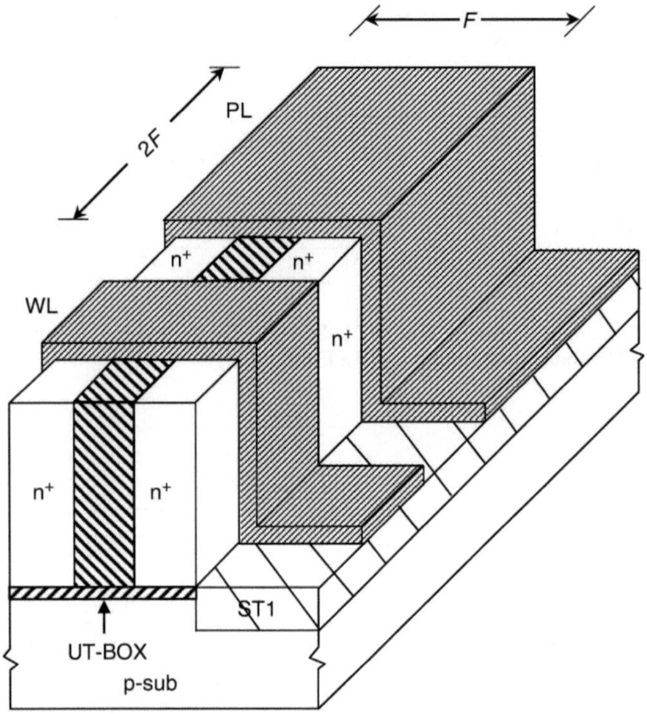

Fig. 5.43 FinFET DRAM cell. *WL* wordline, *PL* capacitor plate. Reproduced from [26] with permission; © 2010 IEEE

5.8 Subsystems for Wider Margins

Improvement of power supply integrity, reduction in V_{t0} at subsystem level, and low-V_{t0} power switches are particularly important to widen margins at the subsystem level.

5.8.1 Improvement of Power Supply Integrity

Small cores and chips, new architectures such as multicore MPUs, and 3D thermally conscious small-chip integration with high-density through silicon vias (TSVs) [10] enable the development of compact subsystems, which, with their reduced wire-length distributions, ensure power supply integrity throughout the subsystem. The improved power supply integrity makes low-V_{DD} operation possible with reducing the difference between V_{DD} and V_{min} while maintaining wide margins. For such subsystems, drastically reducing the memory array area is particularly important since the array dominates the core or chip. Connecting small cores, each embedding a large-capacity DRAM, with low-resistive global interconnects and meshed power-supply lines, as found in the multidivided array of modern DRAMs [5], enable the achievement of wide-margin high-speed multicore LSIs [56, 57]. For example, a hypothetical 0.5-V 16k-core LSI accommodating as many as 320-Mgate logic and 8-Gb DRAMs on a $10 \times 10\text{-mm}^2$ chip would be feasible in the 11-nm generation although the real challenge is to find applications that can fully utilize such a powerful multicore chip. Each homogeneous core, including 20-kgate logic and 512-Kb DRAM with a $5F^2$ cell [3, 11], would be less than $56 \times 56\ \mu\text{m}^2$.

5.8.2 Reduction in V_{t0} at Subsystem Level

Another key to wider margins at subsystem level is to reduce V_{t0} as much as possible, and to use the low-V_{t0} MOSFET throughout the chip. The value of V_{t0} depends on the tolerable value of the total leakage of the chip that is dictated by the subsystem design. Hence, the more the total leakage is reduced, the lower the V_{t0} for a given total leakage current. For example, if the active area in the chip is confined to the small to minimize the active current, while cutting leakage of the remaining inactive area, as shown in Fig. 5.44, the V_{t0} of the active area can be minimized for a given total leakage. This is because the total current of the chip, which is now equal to the sum of the active current (i.e., $\sum C_k V_{DD} f$, f: frequency) and the leakage current (i.e., $\sum W_k 10^{-V_{t0}/S}$, S: subthreshold swing [5]) of the confined active area, can be larger and the leakage current can be larger too with reduced V_{t0}. Even so, there is a limitation for reducing

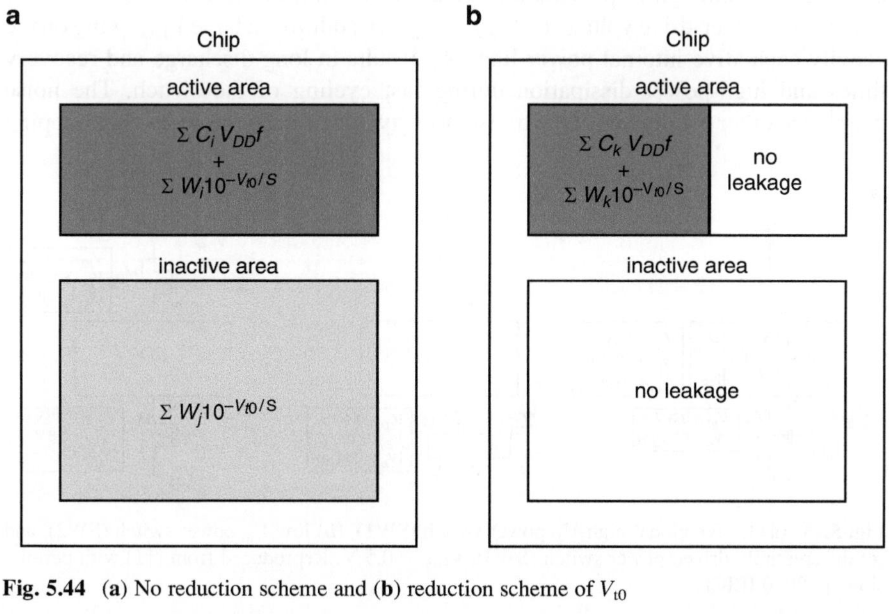

Fig. 5.44 (**a**) No reduction scheme and (**b**) reduction scheme of V_{t0}

the V_{t0}. According to MPU designs, the V_{t0} has been reduced to the point where the leakage occupies about 20% of the total active current.

Various leakage reduction schemes [4, 5], as discussed earlier, are effective to reduce leakage currents in such subsystems. In addition, power switches are indispensable to cut the leakage of the inactive area. However, it always entails large voltage swings on heavily capacitive internal power lines, large spike currents, noise to other conductors, and thus slow recovery time at every switching, which may cause subsystem errors. In addition, it needs an excessively large MOSFET to provide a larger enough active current to the load (i.e., core), although the channel width must be much less than the total width of the internal MOSFETs to minimize the area overhead. Furthermore, there may be challenges to low-V_{DD} and high-speed switching, especially high-speed core-to-core hopping for more advanced multicore LSIs.

5.8.3 Low-V_{t0} Power Switches

Low-V_{t0} power switches are indispensable to overcome above-described drawbacks involved in the conventional power switch. Figure 5.45 compares three power switches [11] designed for application to an internal low-V_{t0} core operating at $V_{DD} = 0.5$ V. To maximize the leakage reduction with the body bias effects [5], the substrates of the P- and N-MOSFETs in the core are connected to V_{DD} and V_{SS}, respectively. The switch in (a) is a conventional high-V_{t0} NMOS (M_S) power switch (SW1), the gate of which is driven at V_{DD} swing. In the inactive mode (i.e., power shut down mode), it completely cuts the core leakage. The channel width, $W(M_S)$, must be wide enough to provide a large active current to the core because of the reduced gate over-drive voltage ($=V_{DD} - V_{t0}$). In addition, a large V_{DD} swing on the heavily capacitive internal power line, N_S, results in long discharge and recovery times and high-power dissipation during fast cycling of the switch. The noise coupled to other conductors at the transients may increase. Fast core-to-core hopping

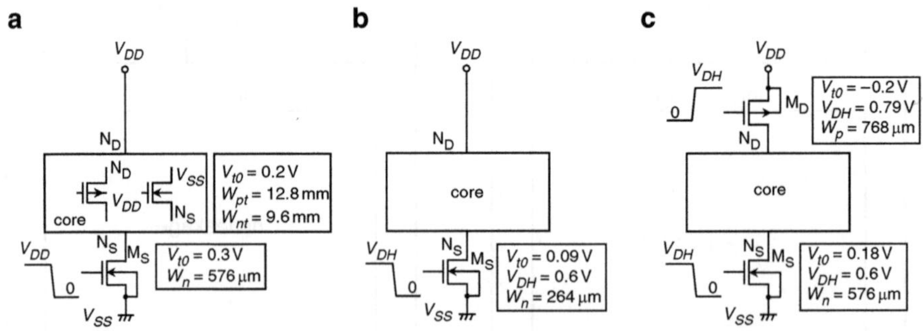

Fig. 5.45 (**a**) Conventional high-V_{t0} power switch (SW1), (**b**) low-V_{t0} power switch (SW2), and (**c**) differentially driven power switch (SW3). $V_{DD} = 0.5$ V. Reproduced from [11] with permission; © 2010 IEICE

is thus prevented. Moreover, each node loses its logic state because it is completely discharged, meaning that a data latch is needed at the node in some cases.

The second switch (b) is a low-V_{t0} NMOS (M_S) power switch (SW2). In the inactive mode, the N_S voltage, V_{NS}, is adjusted so that the total leakage from all the N- and P-MOSFETs in the core is reduced to the value of the current of the leaky (i.e., low V_{t0}) M_S at $V_{GS} = 0$. This reduction stems from the body bias effects and leakage characteristics of MOSFETs. For the NMOSFETs in the core, the leakage is reduced as a result of increasing V_{t0} by raising V_{NS}. The supply voltage of the core is thus reduced to V_{DD}–V_{NS}. Note that the reduced supply voltage is simply the drain-source voltage, V_{DS}, of the switched-off PMOSFETs. The leakage of the PMOSFETs is thus reduced since it is reduced with the reduction in V_{DS} unless V_{DS} is sufficiently high [4]. In any event, M_S can supply

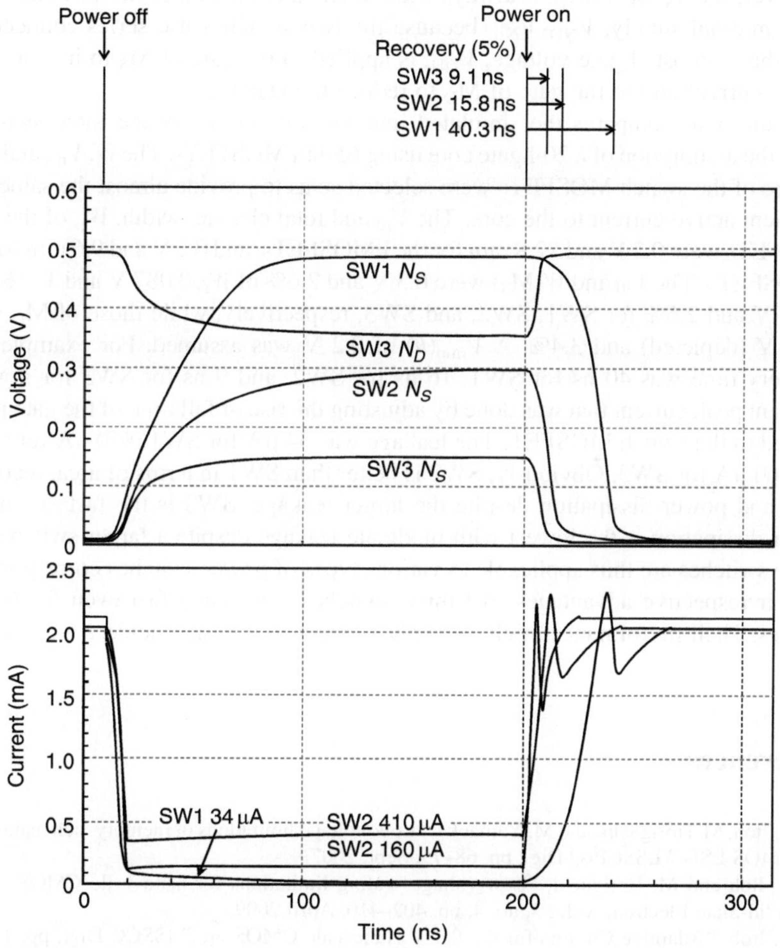

Fig. 5.46 Simulated internal waveforms of a 65-nm 20-kgate core. Reproduced from [11] with permission; © 2010 IEICE

more current to the core in the active mode due to the low V_{t0}, or $W(M_S)$ can be smaller for a given supply current. Moreover, discharge and recovery times and power dissipation are reduced owing to reduced voltage swing. If $V_{DD} - V_{NS} > V_{min}(h)$, the logic state is held, where $V_{min}(h)$ is the minimum supply voltage necessary to hold the logic state.

The third switch (c) is a differentially driven low-V_{t0} PMOS/NMOS (M_D, M_S) power switch (SW3). Leakage for both the N- and P-MOSFETs is reduced due to the body bias effects more than for SW2. The differential operations of the internal N_D and N_S power lines at reduced swing results in short discharge and recovery times, LP dissipation, and low noise. Differential operation occurs when the off-currents of M_S and M_D are equal. The internal logic state is preserved if the internal supply voltage (i.e., voltage difference between N_D and N_S) is higher than $V_{min}(h)$. However, $W(M_S)$ or $W(M_D)$ is always wider than that of the other two switches for a given internal supply, V_{ND}–V_{NS}, because the two switches use series connection. Note that a boosted gate voltage, V_{DH}, is applied to the gate of M_S to increase the supply current and to the gate of M_D to reduce the leakage.

Figure 5.46 compares the simulated internal waveforms for the three switches under the assumption of a 20-kgate core using 65-nm MOSFETs. The W, V_{t0}, and gate voltage of the switch MOSFETs were selected so as to provide almost the same and sufficient active current to the core. The V_{t0} and total channel width, W_t, of the core MOSFETs were 0.2 V and 12.8 mm for the PMOSFETs, and 0.2 V and 9.6 mm for the NMOSFETs. The V_{t0} and $W(M_S)$ were 0.3 V and 2.6% of W_t, 0.087 V and 1.2%, and 0.178 V and 2.6% for SW1, SW2, and SW3, respectively, while those of M_D were −0.2 V (depleted) and 3.4%. A $V_{min}(h)$ of 0.2 V was assumed. For example, the recovery time was 40 ns for SW1, 16 ns for SW2, and 9 ns for SW3 for a given transient peak current that was done by adjusting the rise or fall time of the gate pulse applied to the switch MOSFET. The leakage was 34 μA for SW1, 410 μA for SW2, and 160 μA for SW3. Obviously, SW2 is better than SW1 in terms of area, recovery time, and power dissipation despite the larger leakage. SW3 is the fastest, and its power dissipation is the lowest with moderate leakage despite a larger switch area. These switches are thus applicable to various types of power switches corresponding to their respective advantages. All three switches were quite fast even for 60-nm devices. Their performance might be further enhanced if 11-nm devices were used.

References

1. K. Itoh, M. Horiguchi, and M. Yamaoka, "Low-voltage limitations of memory-rich nano-scale CMOS LSIs," ESSCIRC Dig., pp. 68–75, Sept. 2007.
2. K. Itoh and M. Horiguchi, "Low-voltage scaling limitations for nano-scale CMOS LSIs," Solid-State Electron., vol. 53, no. 4, pp. 402–410, April 2009.
3. K. Itoh, "Adaptive Circuits for the 0.5-V Nanoscale CMOS era," ISSCC Dig., pp. 14–20, Feb. 2009.
4. K. Itoh, M. Horiguchi, and H. Tanaka, *Ultra-Low Voltage Nano-Scale Memories*, Springer, New York, 2007.

5. K. Itoh, *VLSI Memory Chip Design*, Springer, New York, 2001.
6. Y. Nakagome, M. Horiguchi, T. Kawahara, and K. Itoh, "Review and prospects of low-voltage RAM circuits," IBM J. Res. Dev., vol. 47, no. 5/6, pp. 525–552, Sep./Nov. 2003.
7. J.A. Davis, R. Venkatesan, A. Kaloyeros, M. Beylansky, S.J. Souri, K. Banerjee, K.C. Saraswat, A. Rahman, R. Reif, and J.D. Meindl, "Interconnect Limits on Gigascale Integration (GSI) in the 21st Century," Proc. IEEE, vol. 89, no. 3, pp. 305–324, March 2001.
8. W. Haensch, E.J. Nowak, R.H. Dennard, P.M. Solomon, A. Bryant, O.H. Dokumaci, A. Kumar, X. Wang, J.B. Johnson, and M.V. Fischetti, "Silicon CMOS devices beyond scaling," IBM J. Res. Dev., vol. 50, no. 4/5, pp. 339–361, July/Sept. 2006.
9. T.C. Chen, "Where CMOS is going: trendy hype vs. real technology," ISSCC Dig., pp. 22–28, Feb. 2006.
10. A.W. Topol, D.C. La Tulipe Jr., L. Shi, D.J. Frank, K. Bernstein, S.E. Steen, A. Kumar, G.U. Singco, A.M. Young, K.W. Guarini, and M. Ieong, "Three-dimensional integrated circuits," IBM J. Res. Dev., vol. 50, no. 4/5, pp. 491–506, July/Sept. 2006.
11. K. Itoh, M. Yamaoka, and T. Oshima, "Adaptive Circuits for the 0.5-V Nanoscale CMOS Era," IEICE, vol. E93-C, no. 3, pp. 216–233, March 2010.
12. M.J.M. Pelgrom and A.C.J. Duinmaijer, "Matching properties of MOS transistors," J. SSC, vol. 24, no. 5, pp. 1433–1439, Oct. 1989.
13. K. Takeuchi, T. Fukai, T. Tsunomura, A.T. Putra, A. Nishida, S. Kamohara, and T. Hiramoto, "Understanding random threshold voltage fluctuation by comparing multiple fabs and technologies," IEDM Dig., pp. 467–470, Dec. 2007.
14. M. Yamaoka, N. Maeda, Y. Shinozaki, Y. Shimazaki, K. Nii, S. Shimada, K. Yanagisawa, and T. Kawahara, "Low-power embedded SRAM modules with expanded margins for writing," ISSCC Dig., pp. 480–481, Feb. 2005.
15. K. Zhang, U. Bhattacharya, Z. Chen, F. Hamzaoglu, D. Murray, N. Vallepalli, Y. Wang, B. Zheng, and M. Bohr, "A 3-GHz 70MB SRAM in 65 nm CMOS technology with integrated column-based dynamic power supply," ISSCC Dig., pp. 474–475, Feb. 2005.
16. M. Yabuuchi, K. Nii, Y. Tsukamoto, S. Ohbayashi, S. Imaoka, H. Makino, Y. Yamagami, S. Ishikura, T. Terano, T. Oashi, K. Hashimoto, A. Sebe, G. Okazaki, K. Satomi, H. Akamatsu, and H. Shinohara, "A 45 nm low-standby-power embedded sram with improved immunity against process and temperature variations," ISSCC Dig., pp. 326–327, Feb. 2007.
17. K. Itoh, K. Osada, and T. Kawahara, "Reviews and future prospects of low-voltage embedded RAMs," CICC2004 Dig., pp. 339–344, Oct. 2004.
18. H. Masuda, S. Okawa, and M. Aoki, "Approach for physical design in sub-100 nm era," ISCAS Dig., vol. 6, pp. 5934–5937, May 2005.
19. S. Mukhopadhyay, K. Kim, K.A. Jenkins, C-T Chuang, and K. Roy, "Statistical characterization and on-chip measurement methods for local random variability of a process using sense-amplifier-based test structure," ISSCC Dig., pp. 400–401, Feb. 2007.
20. K.J. Kuhn, "Reducing variation in advanced logic technologies: approaches to process and design for manufacturability of nanoscale CMOS," IEDM Dig., pp. 471–474, Dec. 2007.
21. Y. Morita, R. Tsuchiya, T. Ishigaki, N. Sugii, T. Iwamatsu, T. Ipposhi, H. Oda, Y. Inoue, K. Torii, and S. Kimura, "Smallest V^{th} variability achieved by intrinsic thin channel on thin BOX (SOTB) CMOS with single metal gate," Symp. VLSI Tech. Dig., pp. 166–167, June 2008.
22. O. Weber, O. Faynot, F. Andrieu, C. Buj-Dufournet, F. Allain, P. Scheiblin, J. Foucher, N. Daval, D. Lafond, L. Tosti, L. Brevard, O. Rozeau, C. Fenouillet-Beranger, M. Marin, F. Boeuf, D. Delprat, K. Bourdelle, B.-Y. Nguyen, and S. Deleonibus, "High immunity to threshold voltage variability in undoped ultra-thin FDSOI MOSFETs and its physical understanding," IEDM Dig., 10.4, pp. 245–248, Dec. 2008.
23. D. Hisamoto, T. Kaga, Y. Kawamoto, and E. Takeda, "A fully depleted lean-channel transistor (DELTA) – a novel vertical ultra thin SOI MOSFET," IEDM Dig., pp. 833–836, Dec. 1989.
24. R. Tsuchiya, M. Horiuchi, S. Kimura, M. Yamaoka, T. Kawahara, S. Maegawa, T. Ipposhi, Y. Ohji, and H. Matsuoka, "Silicon on thin BOX: a new paradigm of the CMOSFET for

low-power and high-performance application featuring wide-range back-bias control," IEDM Dig., pp. 631–634, Dec. 2004.

25. R. Tsuchiya, N. Sugii, T. Ishigaki, Y. Morita, H. Yoshimoto, K. Torii, and S. Kimura, "Low voltage (V_{dd}~0.6 V) SRAM operation achieved by reduced threshold voltage variability in SOTB (silicon on thin BOX)," VLSI 2009 Tech. Dig., pp. 150–151, June 2009.
26. K. Itoh, N. Sugii, D. Hisamoto, and R. Tsuchiya, "FD-SOI MOSFETs for the low-voltage nanoscale CMOS era," Int. SOI Conference Dig., Oct. 2009.
27. N. Sugii, R. Tsuchiya, T. Ishigaki, Y. Morita, H. Yoshimoto, K. Torii, and S. Kimura, "Comprehensive study on Vth variability in silicon on thin BOX(SOTB) CMOS with small random-dopant fluctuation: finding a way to further reduce variation," IEDM Dig., pp. 249–252, Dec. 2008.
28. K. Itoh and R. Tsuchiya, "Voltage scaling limitations and challenges of memory-rich nanoscale CMO LSIs," MIEL2010 Dig., pp. 39–43, May 2010.
29. S.M. Kim, E.J. Yoon, H.J. Jo, M. Li, C.W. Oh, S.Y. Lee, K.H. Yeo, M.S. Kim, S.H. Kim, D.U. Choe, J.D. Choe, S.D. Suk, D-W Kim, D. Park, K. Kim, and B-Il Ryu, "A novel multi-channel field effect transistor (McFET) on bulk Si for high performance sub-80nm application," IEDM Dig., pp. 639–642, Dec. 2004.
30. H. Sunami, "The role of the trench capacitor in DRAM innovation," IEEE SSCS News, Winter 2008, pp. 42–44, Jan. 22, 2008.
31. J. Amon, A. Kieslich, L. Heineck, T. Schuster, J. Faul, J. Luetzen, C. Fan, C.C. Huang, B. Fischer, G. Enders, S. Kudelka, U. Schroeder, K.H. Kuesters, G. Lange, and J. Alsmeier, "A highly manufacturable deep trench based DRAM cell layout with a planar array device in a 70 nm technology," IEDM Dig., pp. 73–76, Dec. 2004.
32. Y. Nakagome, K. Itoh, M. Isoda, K. Takeuchi, and M. Aoki, "Sub-1-V swing bus architecture for future low-power ULSIs," Symp. VLSI Circuits Dig., pp. 82–83, June 1992.
33. K. Itoh, M. Yamaoka, and T. Kawahara, "Low-voltage limitations of deep-sub-100-nm CMOS LSIs-view of memory designers," GLSVLSI2007 Proc., pp. 529–533, March 2007.
34. T. Tanzawa, A. Umezawa, M. Kuriyama, T. Taura, H. Banba, T. Miyabe, H. Shiga, Y. Takano, and S. Atsumi, "Wordline Voltage Generation System for Low-Power Low-Voltage Flash Memories," IEEE J. Solid-State Circuits, vol. 36, no. 1, pp. 55–63, Jan. 2001.
35. T.I. Chappell, B.A. Chappell, S.E. Schuster, J.W. Allan, S.P. Klepner, R.V. Joshi, and R.L. Franch, "A 2-ns Cycle, 3.8-ns Access 512Kb CMOS ECL SRAM with a fully pipelined architecture," IEEE JSSC, pp. 1577–1584, Nov. 1991.
36. T. Mori, B. Amrutur, K. Mai, M. Horowitz, I. Fukushi, T. Izawa, and S. Mitarai, "A 1 V 0.9 mW at 100 MHz 2x16b SRAM utilizing a half-swing pulsed-decoder and write-bus architecture in 0.25 μm dual-Vt CMOS," ISSCC Dig., pp. 354–355, Feb. 1998.
37. G. Braceras, A. Roberts, J. Conner, R. Wistort, T. Frederick, M. Robillard, S. Hall, S. Burns, and M. Graf, "A 940MHz data-rate 8Mb CMOS SRAM," ISSCC Dig., pp. 198–199, Feb. 1999.
38. T. Kirihata, G. Mueller, B. Ji, G. Frankowsky, J. Ross, H. Terletzki, D. Netis, O. Weinfurtner, D. Hanson, G. Daniel, L. Hsu, D. Storaska, A. Reith, M. Hug, K. Guay, M. Selz, P. Poechmueller, H. Hoenigschmid, and M. Wordeman, "A 390 mm^2 16 bank 1Gb DDR SDRAM with hybrid bitline architecture," ISSCC Dig., pp. 422–423, Feb. 1999.
39. N. Verma and A.P. Chandrakasan, "A 65nm 8T sub-Vt SRAM employing sense-amplifier redundancy," ISSCC Dig., pp. 328–329, Feb. 2007.
40. M. Yamaoka, K. Osada, and T. Kawahara, "A cell-activation-time controlled SRAM for low-voltage operation in DVFS SoCs using dynamic stability analysis," ESSCIRC Dig., pp. 286–289, Sept. 2008.
41. M. Khellah, N.S. Kim, J. Howard, G. Ruhl, M. Sunna, Y. Ye, J. Tschanz, D. Somasekhar, N. Borkar, F. Hamzaoglu, G. Pandya, A. Farhang, K. Zhang, and V. De, "A 4.2GHz 0.3 mm^2 256kb Dual-Vcc SRAM building block in 65nm CMOS," ISSCC Dig., pp. 624–625, Feb. 2006.
42. K. Itoh, A.R. Fridi, A. Bellaouar, and M.I. Elmasry, "A deep sub-V, single power-supply SRAM cell with multi-V_T, boosted storage node and dynamic load," Symp. VLSI Circuits Dig., pp. 132–133, June 1996.

43. J. Pille, C. Adams, T. Christensen, S. Cottier, S. Ehrenreich, F. Kono, D. Nelson, O. Takahashi, S. Tokito, O. Torreiter, O. Wagner, and D. Wendel, "Implementation of the CELL broadband engine in a 65 nm SOI technology featuring dual-supply SRAM arrays supporting 6GHz at 1.3V," ISSCC Dig., pp. 322–323, 606, Feb. 2007.
44. H. Akamatsu, T. Iwata, H. Yamamoto, T. Hirata, H. Yamauchi, H. Kotani, and A. Matsuzawa, "A low power data holding circuit with an intermittent power supply scheme for sub-1V MT-CMOS LSIs," Symp. VLSI Circuits Dig., pp. 14–15, June 1996.
45. M. Yamaoka, Y. Shinozaki, N. Maeda, Y. Shimazaki, K. kato, S. Shimada, K. Yanagisawa, and K. Osada, "A 300 MHz 25 μA/Mb leakage on-chip SRAM module featuring process-variation immunity and low-leakage-active mode for mobile-phone application processor," ISSCC Dig., pp. 494–495, Feb. 2004.
46. F. Hamzaoglu, K. Zhang, Y. Wang, H.J. Ann, U. Bhattacharya, Z. Chen, Y-G Ng, A. Pavlov, K. Smits, and M. Bohr, "A 153Mb-SRAM design with dynamic stability enhancement and leakage reduction in 45 nm high-k metal-gate CMOS technology," ISSCC Dig., pp. 376–377, Feb. 2008.
47. H. Pilo, V. Ramadurai, G. Braceras, J. Gabric, S. Lamphier, and Y. Tan, "A 450ps access-time SRAM macro in 45 nm SOI featuring a two-stage sensing scheme and dynamic power management," ISSCC Dig., pp. 378–379, Feb. 2008.
48. K. Nii, M. Yabuuchi, Y. Tsukamoto, S. Ohbayashi, Y. Oda, K. Usui, T. Kawamura, N. Tsuboi, T. Iwasaki, K. Hashimoto, H. Makino, and H. Shinohara, "A 45-nm single-port and dual-port SRAM family with robust read/write stabilizing circuitry under DVFS environment," Symp. VLSI Circuits Dig., pp. 212–213, June 2008.
49. H. Nho, P. Kolar, F. Hamzaoglu, Y. Wang, E. Karl, Y-G. Ng, U. Bhattacharya, and K. Zhang, "A 32 nm high-k metal gate SRAM with adaptive dynamic stability enhancement for low-voltage operation," ISSCC Dig., pp. 346–347, Feb. 2010.
50. L. Chang, Y. Nakamura, R.K. Montoye, J. Sawada, A.K. Martin, K. Kinoshita, F.H. Gebara, K.B. Agarwal, D.J. Acharyya, W. Haensch, K. Hosokawa, and D. Jamsek, "A 5.3 GHz 8T-SRAM with operation down to 0.41 V in 65 nm CMOS," Symp. VLSI Circuits Dig., pp. 252–253, 2007.
51. S. Akiyama, T. Sekiguchi, R. Takemura, A. Kotabe, and K. Itoh, "Low-V_T small-offset gated preamplifier for sub-1-V gigabit DRAM arrays," ISSCC Dig., pp. 142–143, Feb. 2009.
52. A. Kotabe, Y. Yanagawa, S. Akiyama, and T. Sekiguchi, "CMOS low-V_T preamplifier for 0.5-V gigabit-DRAM array," A-SSCC2009 Dig., pp. 213–216, Nov. 2009.
53. A. Kotabe, Y. Yanagawa, R.Takemura, T. Sekiguchi, and K. Itoh, "Asymmetric cross-coupled sense amplifier for small-sized 0.5-V gigabit-DRAM arrays," to be presented at A-SSCC2010, Nov. 2010.
54. H. Gotou, Y. Arimoto, M. Ozeki, and K. Imaoka, "Soft error rate of SOI-DRAM," IEDM Dig., pp. 870–871, 1987.
55. R. Takemura, K. Itoh, and T. Sekiguchi, "A 0.5-V FD-SOI Twin-Cell DRAM with offset-free dynamic-VT sense amplifier," ISLPED Dig., pp. 123–126, Oct. 2006.
56. D. Truong, W. Cheng, T. Mohsenin, Z. Yu, T. Jacobson, G. Landge, M. Meeuwsen, C. Watnik, P. Mejia, A. Tran, J. Webb, E. Work, Z. Xiao, and B. Baas, "A 167-processor 65 nm computational platform with per-processor dynamic supply voltage and dynamic clock frequency scaling," Symp. VLSI Circuits Dig., pp. 22–23, June 2008.
57. S. Vangal, J. Howard, G. Ruhl, S. Dighe, H. Wilson, J. Tschanz, D. Finan, P. Iyer, A. Singh, T. Jacob, S. Jain, S. Venkataraman, Y. Hoskote, and N. Borkar, "An 80-tile 1.28TFLOPS network-on-chip in 65 nm CMOS," ISSCC Dig., pp. 98–99, Feb. 2007.

Chapter 6
Reduction Techniques for Speed-Relevant Errors of Nanoscale Memories

6.1 Introduction

There are two kinds of speed-relevant errors, regarded as emerging problems in the low-voltage nanoscale CMOS era, which are the speed-degradation error and the interdie speed-variation error. The errors become more prominent as operating voltage V_{DD} approaches the unscalable and high value of the lowest necessary average V_t (i.e., $V_{t0} = 0.2 - 0.4$ V, see Fig. 5.5). The speed-degradation error can occur for an average chip in a wafer whenever an average MOSFET in the chip slows down with reduced gate-over-drive voltage ($= V_{DD} - V_{t0}$), and the resultant speed of the chip does not meet the speed target. Note that even in the absence of any variation (i.e., interdie and intradie variations), all chips can be faulty with slow speed. The solutions are to develop new higher speed designs and higher speed MOSFETs to offset reduced gate-over-drive voltage, as described later.

The interdie speed-variation error statistically occurs for some chips in a wafer when the speed distribution of chips broadens due to the so-called global variation of design parameters in the wafer (e.g., V_t and channel-length variations) that mainly comes from V_t lowering by short-channel effects. Even if the speed of the average chip is much faster than the target, this error happens as V_{DD} approaches to V_{t0}. Figure 6.1 shows speed distributions of circuits in a chip and chips in a wafer. If a chip comprises many identical circuits, as in a memory cell array, speed variations occur for the circuits due to the local variation mainly coming from random dopant fluctuation (RDF), as discussed in Chap. 5, showing a speed distribution with the average speed τ_0' at the average V_t of the chip (i.e., V_{t0}'). The interdie speed variation makes the average speed τ_0' of each chip distribute with the average speed τ_0 at the average V_t (i.e., V_{t0}) of the wafer, as shown in the figure. Even if the speed of the average chip in the wafer meets the speed target, some of the chips result in faults due to a speed spread of chips widened by reducing V_{DD}. A solution is power management [1–3] for compensating for the interdie speed variation with static or quasistatic controls of internal supply voltages. Substrate voltage (V_{BB}) control with an on-chip V_{BB} generator, widely used for stable operation of NMOS DRAMs in the 1980s and 1990s, may be applicable to reduce the speed variation of chips, as discussed later. Internal V_{DD} control with an on-chip V_{DD} generator (i.e., on-chip

M. Horiguchi and K. Itoh, *Nanoscale Memory Repair*, Integrated Circuits and Systems 1,
DOI 10.1007/978-1-4419-7958-2_6,

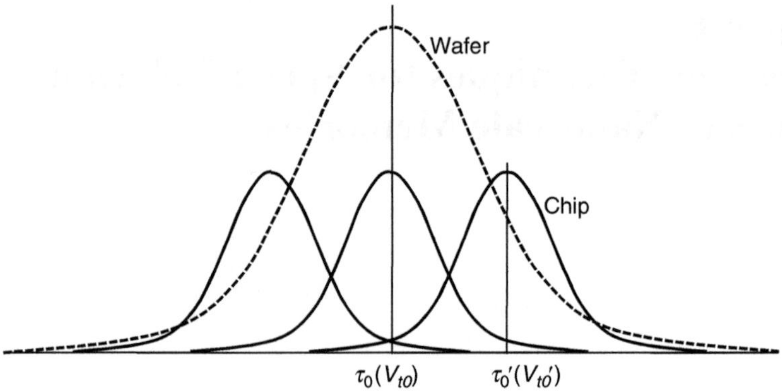

Fig. 6.1 Speed distributions caused by intradie and interdie V_t variations

voltage down converter) may play a key role in compensation for other interdie parameter variations that the V_{BB} control cannot manage. Furthermore, combination of the V_{BB} and V_{DD} controls may be realistic to tackle the variation issue in the nanoscale era.

In this chapter, power-management problems expected in the nanoscale memories, and possible solutions are mainly described.

6.2 Reduction Techniques for Speed-Degradation Errors

If V_{min} is reduced sufficiently under a given V_{DD}, the resultant wide voltage margin allows V_{DD} to be reduced, giving a benefit of low-power dissipation. This reduction, however, is strictly limited by speed errors that occur for an average chip. Figure 6.2 plots the speed of MOSFET [see (1.1)] normalized with that of $V_{DD} = 1.2$ V. For example, for an average V_t (i.e., V_{t0}) $= 0.35$ V, speed is degraded to about one-third when V_{DD} is reduced from 1.2 to 0.5 V. If this degradation is intolerable, V_{DD} cannot be reduced to 0.5 V even if V_{min} is less than 0.5 V. There are two ways to cope with the problem. One is to effectively reduce V_{t0} by using low-V_{t0} circuits throughout the chip, as suggested from characteristics for $V_{t0} = 0$ V in the figure. In addition to circuits discussed in Chap. 5, the well-known dynamic V_t circuit may be a solution, in which V_t is changed from a high value for off-MOSFET to a low value for on-MOSFET by dynamic controls of the substrate of an ultrathin BOX double-gate FD-SOI [3]. The other is to enlarge the channel width W, so the degraded gate-over-drive voltage is offset. This may be achieved without incurring area penalty with vertical MOSFETs such as FinFETs, although it is ineffective due to increased power dissipation when the load capacitance is dominated by MOS gate.

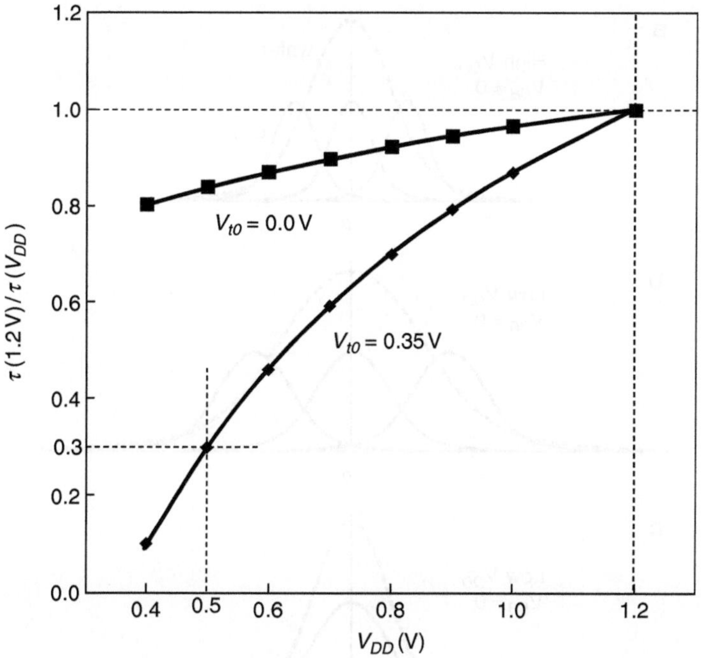

Fig. 6.2 Speed degradation with reducing V_{DD}

6.3 Reduction Techniques for Interdie Speed-Variation Errors

Unfortunately, the distribution of interdie speed variation broadens with reducing V_{DD}, as shown in Fig. 6.3a, b, resulting in increase in faulty chips. Then, let us investigate how it broadens on the assumption that each chip in a wafer is composed of many identical MOSFETs with interdie variations in the channel length L and threshold voltage V_t. In this case, the speed ratio $\Delta\tau$ of the slowest chip, composed of the MOSFET with the largest channel length L (i.e., $L_0 + \Delta L_{max}$) and the highest V_t (i.e., $V_{t0} + \Delta V_{tmax}$), to the average chip, composed of an average MOSFET with average values in L (i.e., L_0) and V_t (i.e., V_{t0}), is approximately given as

$$\begin{aligned}\Delta\tau &= \tau(V_{t0} + \Delta V_{tmax}, L_0 + \Delta L_{max})/\tau(V_{t0}, L_0)\\ &= \{(L_0 + \Delta L_{max})/L_0\}\{1 - \Delta V_{tmax}/(V_{DD} - V_{t0})\}^{-1.2}.\end{aligned} \quad (6.1)$$

Figure 6.4 plots $\Delta\tau$ vs. V_{DD} for $(L_0 + \Delta L_{max})/L_0 = 1.1$, $\Delta V_{tmax} = 50$ mV, and $V_{t0} = 0.35$ V. It is obvious that $\Delta\tau$ becomes intolerably large as V_{DD} approaches to the V_{t0} ($\cong$0.35 V), and may cause faulty chips. For example, $\Delta\tau$ is as small as 1.21 at $V_{DD} = 1$ V, but as large as 1.79 at $V_{DD} = 0.5$ V. Compensation techniques for

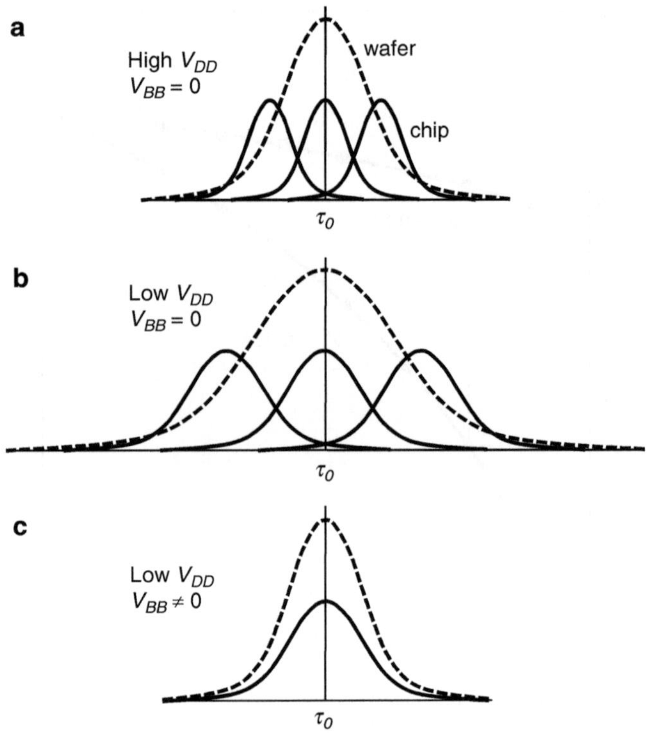

Fig. 6.3 Distribution of speed variation of chips and compensation with V_{BB} application. The distribution is widened by applying (**a**) a high V_{DD} and (**b**) a low V_{DD}. All chips are changed so as to have a unified speed by (**c**) V_{BB} application

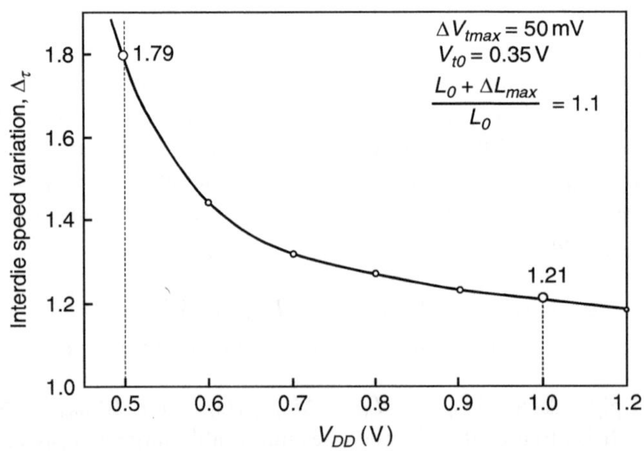

Fig. 6.4 Interdie speed variation vs. V_{DD}

the variations are thus indispensable, especially for lower voltage operations. An effective solution is to compensate with power management, that is, adaptive controls of internal power supplies, such as substrate bias V_{BB} and operating voltage V_{DD}, as described below.

6.3.1 On-Chip V_{BB} Compensation

If V_{t0} is adjustable with V_{BB}, the speed variation of the chip caused by interdie V_t variations is compensated for by controlling V_{BB}. Hence, ideally, all chips have a unified speed, as seen in Fig. 6.3c. Here, an internal V_{BB} control with an on-chip V_{BB} generator [1–3] is indispensable for general purpose LSIs, because an external V_{BB} control needs an additional power supply of V_{BB}. However, careful attention must be paid to the instabilities [1–3] involved in the on-chip V_{BB} control.

In the past, DRAM designers encountered numerous problems that occurred even in static or quasistatic V_{BB} and V_{DD} controls. It is well known that the DRAM has been the only large-volume production LSI using a substrate bias V_{BB} that is supplied from an on-chip V_{BB} generator. In the NMOS DRAM era in the 1970s and 1980s, when a quasistatic V_{BB} was supplied to the p-type substrate of the whole chip (i.e., both array and periphery), the generator caused instabilities (surge current [1] or a degraded V_{BB} level [1]) at power-on and during burn-in high-voltage stress tests, and shortened the refresh time of cells due to minority-carrier injection to cells [1]. Poor current drivability of the generator consisting of charge pumps, a large substrate current generated from MOSFETs in the peripheral circuits, and the substrate structure were mainly responsible for the instabilities. Even so, DRAM designers were fortunate because both the static bias setting of a deep V_{BB} of about -3 V and a sufficiently high V_t of about 0.5 V allowed small changes in V_t, even with quite large quasistatic V_{BB} variations and V_{BB} noise [1, 2], as seen in Fig. 6.5a. Thus, a small ΔV_t and a high V_{t0} result in a small ΔV_t-to-V_{t0} ratio, causing stable operation. In the CMOS DRAM era since the late 1980s, V_{BB} was removed from peripheral circuits primarily to eliminate instabilities caused by the generator and has only been supplied to the array to ensure stable operation. However, in the low-voltage low-V_{t0} CMOS era, in which on-chip V_{BB} compensation will be again needed for the peripheral logic circuits, we may face more serious instability problem than the NMOS DRAM era. Note that the compensation usually relies on strong dependence of V_{t0} on V_{BB}, which is developed at about $V_{BB} = 0$ V, for better compensation efficiency. Instead, operations of low-V_{t0} MOSFETs are more susceptible to V_{BB} noise, causing possible instabilities as follows.

In addition to noise coupled to the V_{BB} line itself, noise coupled to power lines may change V_{t0}. Figure 6.5b illustrates the substrate-noise generation mechanism [1, 2, 4]. A V_{BB} (i.e., V_{BN} for NMOS and V_{BP} for PMOS in the figure) of 0 V means no difference between the substrate and source voltages (i.e., V_{SS} for NMOS and V_{DD} for PMOS). If the substrate line and the source line are separated to enable V_{BB} compensation, a voltage difference (i.e., substrate noise) is actually developed

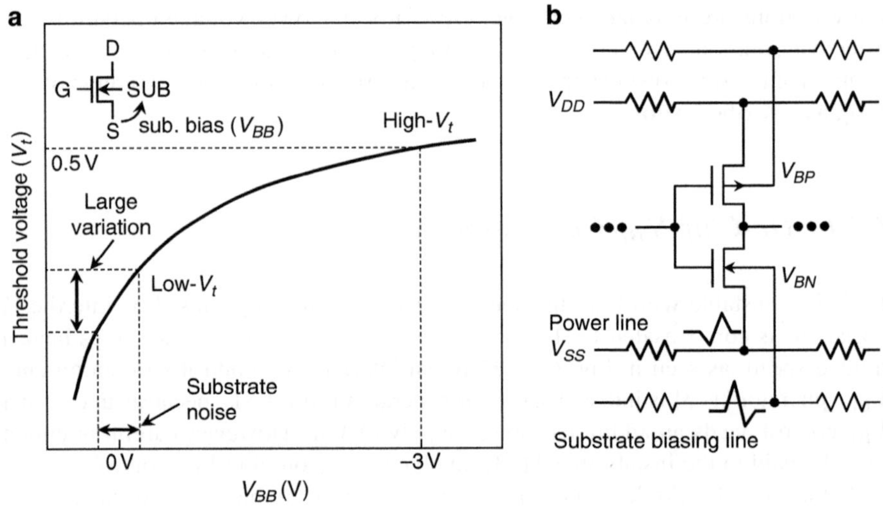

Fig. 6.5 Substrate noise. (**a**) V_t vs. V_{BB} for bulk MOSFETs and (**b**) coupling noise in a CMOS inverter [4]

during high-speed active operation, and thus the voltages coupled to each line can be different. Note peripheral logic circuits in a modern CMOS SRAM/DRAM. As in MPUs and ASICs, the tight connection between the source (i.e., a power line of V_{DD} or V_{SS}) and the well (i.e., substrate) in each MOSFET never creates a voltage difference between source and well, thus ensuring a fixed V_{BB} of 0 V throughout the chip. The voltage difference, however, may cause a large V_t variation, because V_t drops more sharply with a shallower substrate bias, as mentioned earlier. This would result in unstable active operation, especially at low V_{DD} (i.e., low V_{t0}), with a large ΔV_t-to-V_{t0} ratio. To be more exact, in addition to the above-described coupling noise to the substrate and power lines, V_{BB} droop is also hazardous because it is also responsible for the V_t variation. The droop occurs when the total substrate current I_{BB} of MOSFETs exceeds the drive current of the charge pump in the V_{BB} generator. To reduce the droop the pumping current must be increased, or the I_{BB} must be reduced. Unfortunately, the charge pump in the generator has poor current drivability, and it consumes higher power if larger drive current is necessary. Therefore, a smaller I_{BB} MOSFET is desirable for reducing the droop. FD-SOI structures are good candidates due to the negligible I_{BB}, as mentioned below.

Figure 6.6 compares a bulk MOSFET and an ultrathin BOX planar FD-SOI (SOTB, see Fig. 5.10) in terms of compensation capability. An on-chip V_{BB} generator comprises a V_t detector and a charge pump. The charge pump changes V_{BB} according to the detected V_t, so V_t can be set to a desired value. For the bulk, the source/drain-substrate current, I_{BB}, is inherently large due to p–n junctions. The I_{BB} is further enhanced by higher V_{DD} necessary to offset the large V_t variation of the bulk. Hence, V_{BB} tends to be unstable due to the poor current driving capability of

	Bulk	SOTB
Structure	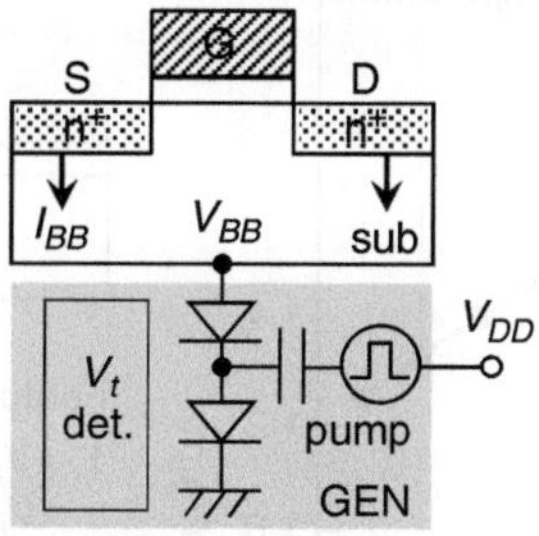	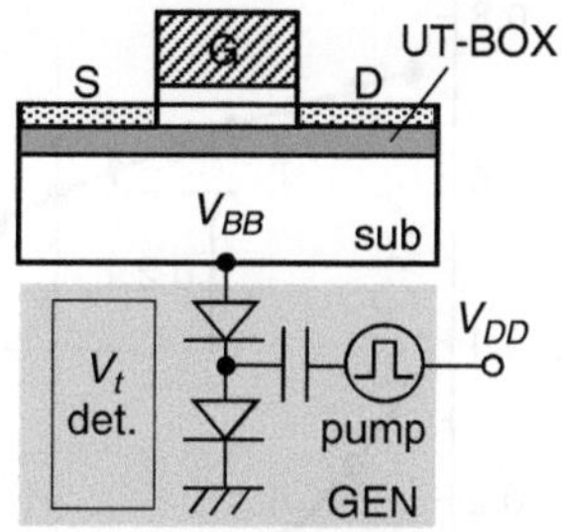
I_{BB}	large → ΔV_{BB} (pn-junc. Higher V_{DD}*)	no → stable V_{BB} (UT-BOX)
On-chip V_{BB}	difficult	possible

$*I_{BB} \propto \exp(-\eta / V_{DD})$

Fig. 6.6 Comparison between a bulk MOSFET and an ultrathin (UT-) BOX planar FD-SOI MOSFET (SOTB) in terms of V_{BB} compensation capability

the charge pump, making the on-chip V_{BB} approach extremely difficult. In contrast, for the SOTB, the UT-BOX stops I_{BB} generation, so the charge pump generates a stable V_{BB}, making the on-chip V_{BB} approach possible. Fortunately, the SOTB realizes a large V_t change of 0.2 V with a V_{BB} change of 1 V without increasing A_{vt}, as shown in Fig. 6.7 [5]. In addition, a positive or negative V_{BB} is applicable without I_{BB} thanks to the UT-BOX structure. Even for FinFETs, V_t is adjustable with V_{BB} if a UT-BOX is used on the bottom. Figure 6.8 shows the device model and simulation results of such a FinFET [6]. Obviously, the potential is sufficiently applied from the substrate to the top of the fin even for a fin height and width of 20 nm. Consequently, a large V_t change of 0.2 V is realized with a V_{BB} change of 1 V for a fin height of 20 nm.

There is a controversy about applying a forward V_{BB} to bulk MOSFETs. Although controlling with forward V_{BB} is more effective in reducing speed variations because the $V_{t0} - V_{BB}$ characteristics are more sensitive to V_{BB}. If a forward V_{BB} must be used, however, requirements to suppress noise become more stringent, calling for a uniform distribution of the forward V_{BB} throughout the chip. Additional current consumption, in the form of bipolar current induced by the forward V_{BB}, is another matter [3] that must be considered.

Special attention must also be paid to instabilities in unusual operating modes. For example, power-on CMOS latch-up and/or the rush current [1] tend to be enhanced by a lower V_{t0} (even for enhancement MOSFET in some cases) that is developed at a shallow V_{BB} of around 0 V during power-on. This is because the heavy substrate capacitance is slowly charged up from a floating 0 V to a floating V_{BB} due to the poor current-driving capability of an on-chip

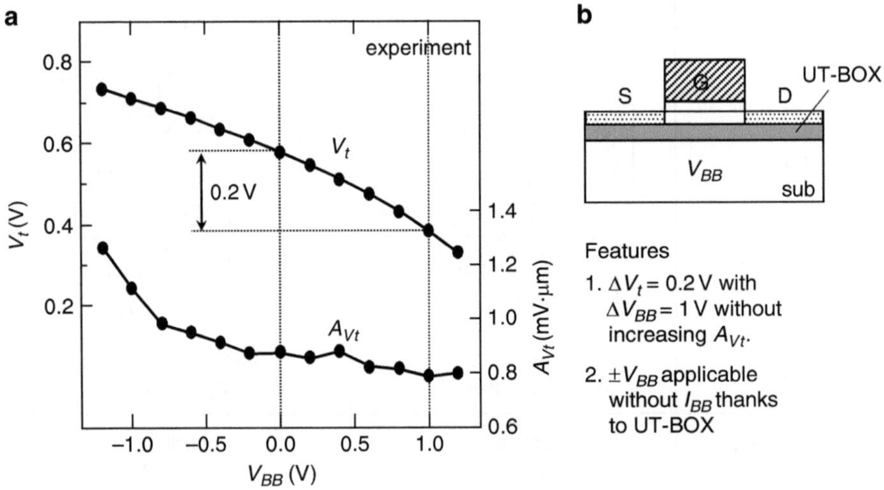

Fig. 6.7 UT-BOX planar FD-SOI MOSFET (SOTB) [5]. (**a**) V_t vs.V_{BB} and (**b**) devise structure and summary of the features

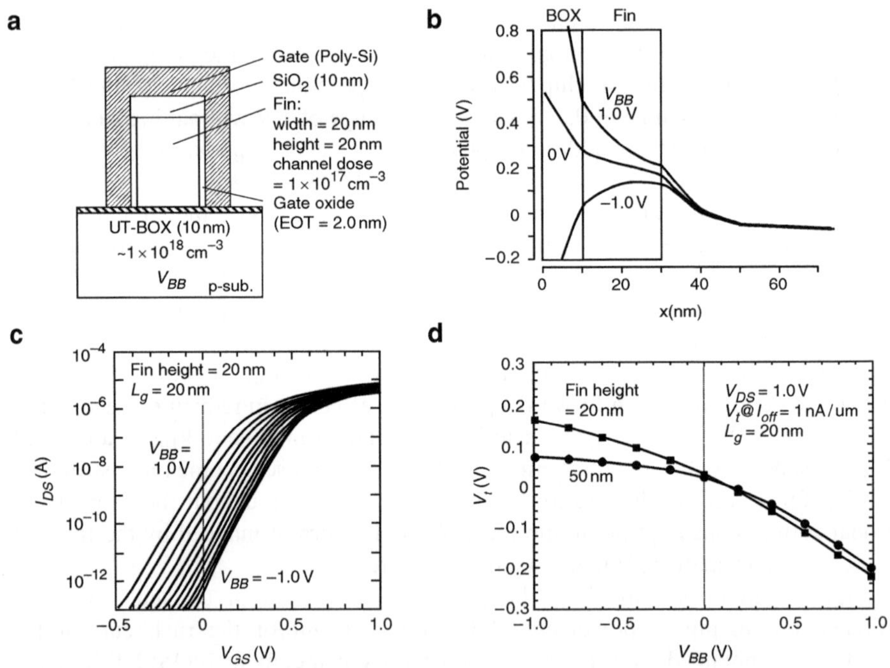

Fig. 6.8 (**a**) FinFET structure for simulation, (**b**) potential along the height, (**c**) I_{DS} vs. V_{GS}, and (**d**) V_t vs. V_{BB} for a fin height of 20 nm. V_t for a fin height of 50 nm is added in (**d**). Reproduced from [6] with permission; © 2010 IEEE

V_{BB} generator. Enhanced instability at burn-in test with a high-stress voltage is another concern [1].

6.3.2 On-Chip V$_{DD}$ Compensation and Others

Internal-V_{DD} control with an on-chip voltage down converter (VDC) [1–3] (i.e., series regulator) may play a key role in compensation for other interdie parameter variations that V_{BB} control cannot manage, as mentioned before. For example, it never reduces subthreshold currents, unlike the V_{BB} control, but it compensates for speed variations more effectively. The VDC design is simpler than the well-known buck converter, and it has been well established in DRAM designs despite lower conversion efficiency. In the low-voltage nanoscale era, VDCs having a low dropout (i.e., a small difference between V_{DD} and the converted voltage V_{DL}, e.g., a dropout < 0.1 V) [1] may be popular for minimizing a power loss at VDCs. Even combination of V_{BB} and V_{DD} controls may be realistic to tackle the variation issue. In fact, such sophisticated compensations are justified by recent experimental data of a 65-nm multicore LSI [7], as shown in Fig. 6.9. The speed variation between 80 cores in a chip turned out to be more serious with reducing V_{DD} even for 65-nm devices. This suggests that even each core needs its own compensation scheme for core-to-core speed variations that are enhanced by further device and voltage scaling. In any event, Fig. 6.10 shows an example of power supplying scheme for nanoscale memories. All internal voltages are adaptively controlled with on-chip voltage converters that operate at an external voltage V_{DD}, so they are compensated for variations of process, voltage, and temperature (PVT variation). Two major voltages are V'_{DD} and V_{DL} for dual-V_{DD} and dual-V_{t0} circuits, respectively. V'_{DD} is a low-dropout voltage from an on-chip

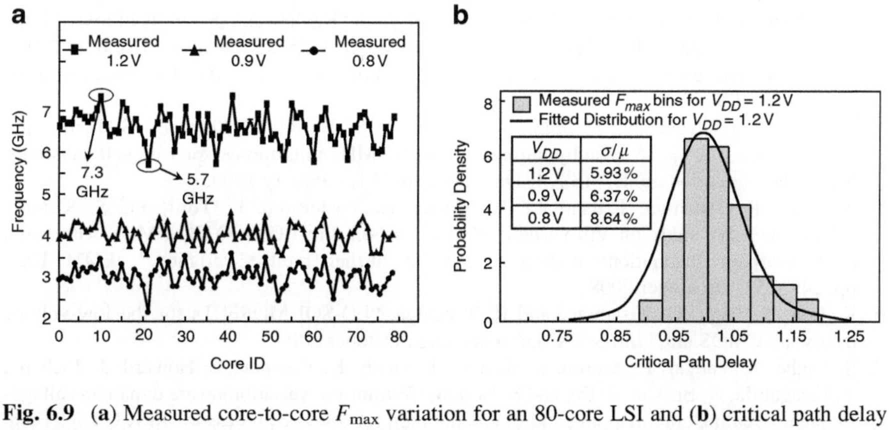

Fig. 6.9 (**a**) Measured core-to-core F_{max} variation for an 80-core LSI and (**b**) critical path delay distribution and its variation, σ/μ, for different V_{DD}. Reproduced from [7] with permission; © 2010 IEEE

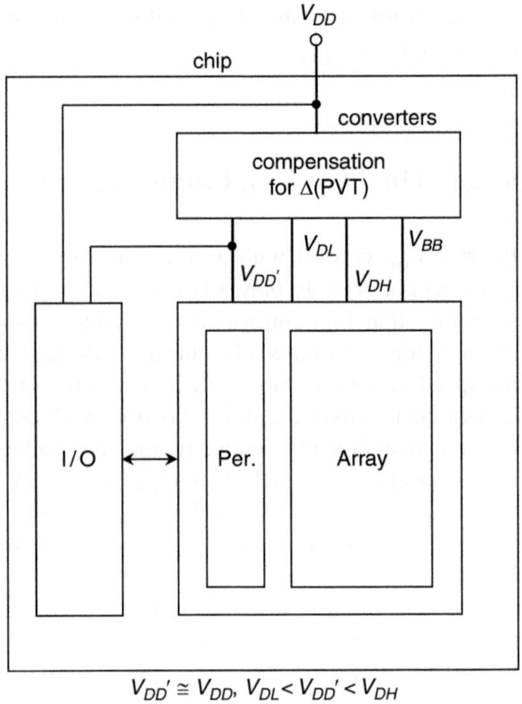

Fig. 6.10 PVT variation-immune nanoscale memory

VDC. V_{BB} is a substrate voltage and V_{DH} is a boosted voltage, for example, for DRAM wordline.

References

1. K. Itoh, *VLSI Memory Chip Design*, Springer, New York, 2001.
2. Y. Nakagome, M. Horiguchi, T. Kawahara, and K. Itoh, "Review and prospects of low-voltage RAM circuits," IBM J. Res. Dev., vol. 47, no. 5/6, pp. 525–552, September/November 2003.
3. K. Itoh, M. Horiguchi, and H. Tanaka, *Ultra-Low Voltage Nano-Scale Memories*, Springer, New York, 2007.
4. H. Mizuno, K. Ishibashi, T. Shimura, T. Hattori, S. Narita, K. Shiozaki, S. Ikeda, and K. Uchiyama, "A 18 μA-standby-current 1.8 V 200 MHz microprocessor with self substrate-biased data retention mode," ISSCC Dig., pp. 280–281, February 1999.
5. N. Sugii, R. Tsuchiya, T. Ishigaki, Y. Morita, H. Yoshimoto, K. Torii, and S. Kimura, "Comprehensive study on Vth variability in silicon on thin BOX(SOTB) CMOS with small random-dopant fluctuation: finding a way to further reduce variation," IEDM Dig., pp. 249–252, December 2008.
6. K. Itoh, N. Sugii, D. Hisamoto, and R. Tsuchiya, "FD-SOI MOSFETs for the low-voltage nanoscale CMOS era," Int. SOI Conference Dig., October 2009.
7. S. Dighe, S. Vangal, P. Aseron, S. Kumar, T. Jacob, K. Bowman, J. Howard, J. Tschanz, V. Erraguntla, N. Borkar, V. De, and S. Borkar, "Within-die variation-aware dynamic-voltage-frequency scaling core mapping and thread hopping for 80-core processor," ISSCC Dig. Tech. Papers, pp. 174–175, February 2010.

Index

M. Horiguchi and K. Itoh, *Nanoscale Memory Repair*, Integrated Circuits and Systems 1,
DOI 10.1007/978-1-4419-7958-2, © Springer Science+Business Media, LLC 2011

MIX
Papier aus verantwortungsvollen Quellen
Paper from responsible sources
FSC® C105338

If you have any concerns about our products,
you can contact us on
ProductSafety@springernature.com

In case Publisher is established outside the EU,
the EU authorized representative is:
Springer Nature Customer Service Center GmbH
Europaplatz 3, 69115 Heidelberg, Germany

Printed by Libri Plureos GmbH
in Hamburg, Germany